Correlations in Diagnostic Imaging

edited by

D. BRUCE SODEE, M.D.

Director, Department of Nuclear Medicine
Nuclear Medicine Institute
Hillcrest Hospital
Mayfield Heights, Ohio

with the assistance of

THOMAS A. VERDON, JR., M.D.

Director, Department of Nuclear Medicine and Ultrasound
The Penrose Hospitals
Colorado Springs, Colorado

CORRELATIONS IN DIAGNOSTIC IMAGING

Nuclear Medicine, Ultrasound, and Computed Tomography in Medical Practice

APPLETON-CENTURY-CROFTS/New York

79 80 81 82 83 / 10 9 8 7 6 5 4 3 2 1

Prentice-Hall International, Inc., London
Prentice-Hall of Australia, Pty. Ltd., Sydney
Prentice-Hall of India Private Limited, New Delhi
Prentice-Hall of Japan, Inc., Tokyo
Prentice-Hall of Southeast Asia (Pte.) Ltd., Singapore
Whitehall Books Ltd., Wellington, New Zealand

Library of Congress Cataloging in Publication Data
Main entry under title:

Correlations in diagnostic imaging.
Bibliography: p.
Includes index.
1. Diagnosis, Radioscopic. 2. Diagnosis, Radioscopic—Cost effectiveness. 3. Imaging systems in medicine.
I. Sodee, D. Brude, 1930– II. Verdon, Thomas A.
RC78.C66 616.07'57 79-4506
ISBN 0-8385-1234-8

Text design: Edmée Froment
Cover design: Judith F. Warm

PRINTED IN THE UNITED STATES OF AMERICA

Contributors

***JASON C. BIRNHOLZ, M.D.**
Assistant Professor of Radiology
Harvard Medical School
Boston, Massachusetts

***ROBERT A. BROOK**
Director, Division of Health Care Systems
Public Health Service
Department of Health, Education, and Welfare
San Francisco, California

PATRICK J. BRYAN, M.D.
Radiologist, St. Vincent's Hospital
Dublin, Ireland

LEE C. CHIU, M.D.
Associate Professor
Radiology Department
Iowa University Hospital
Iowa City, Iowa

WILLIAM M. COHEN, M.D.
Professor of Radiology
State University of New York
The Upstate Medical Center
Syracuse, New York

***ERNEST W. FORDHAM, M.D.**
Chairman, Nuclear Medicine Department
Rush Presbyterian–St. Luke's Medical Center
Chicago, Illinois

GERALD S. FREEDMAN, M.D.
Associate Clinical Professor of Radiology
Yale University School of Medicine
Director, Radiology
Temple Street Medical Center
New Haven, Connecticut

ATIS K. FREIMANIS, M.D.
Professor and Chairman, Radiology Department
Ohio State University College of Medicine
Columbus, Ohio

ZACHARY D. GROSSMAN, M.D.
Associate Professor of Radiology
and Nuclear Medicine
State University of New York
The Upstate Medical Center
Syracuse, New York

* Lecturer or discussion participant

***HIRSCH HANDMAKER, M.D.**
Director, Nuclear Medicine Service
Children's Hospital of San Francisco
San Francisco, California

WALTER L. HENRY, M.D.
Professor of Medicine and
Chief, Division of Cardiology
University of California, Irvine
Irvine, California

***GERALD F. LARK**
Associate Administrator
Hillcrest Hospital
Mayfield Heights, Ohio

GEORGE R. LEOPOLD, M.D.
Professor of Radiology
University of California, San Diego
Chief, Division of Diagnostic Ultrasound
San Diego, California

EDWARD A. LYONS, M.D.
Director, Section of Diagnostic Ultrasound
Associate Professor, Radiology
University of Manitoba
Winnipeg, Canada

KENNETH A. McKUSICK, M.D.
Assistant Professor, Harvard Medical School
Associate Radiologist, Clinical Director
Nuclear Medicine Department
Massachusetts General Hospital
Boston, Massachusetts

RONALD F. MEADOWS
Vice President
Technicare Corporation
Solon, Ohio

FRED S. MISHKIN, M.D.
Director of Nuclear Medicine
Martin Luther King, Jr. General Hospital
Professor of Radiology
Charles R. Drew Postgraduate Medical School
Los Angeles, California

***ROBERT MOSS**
Administrator, Hillcrest Hospital
Mayfield Heights, Ohio

***WILLIAM H. OLDENDORF, M.D.**
Senior Medical Investigator
VA Medical Center—Brentwood
Professor of Neurology and Psychiatry
University of California at Los Angeles School of Medicine
Los Angeles, California

RICHARD J. OSZUSTOWICZ
Assistant Professor, Programs in Hospital and Health-Care Administration
University of Minnesota
Minneapolis, Minnesota

C. LEON PARTAIN, Ph.D., M.D.
Research Associate and Nuclear Medicine Fellow
Imaging Division, Department of Radiology
University of North Carolina
Chapel Hill, North Carolina

GORDON S. PERLMUTTER, M.D.
Clinical Associate Professor of Radiology
Temple University Health Sciences Center
Director of the Ultrasound Laboratory
The Reading Hospital and Medical Center
Reading, Pennsylvania

GERALD M. POHOST, M.D.
Assistant Professor of Medicine, Harvard Medical School
Assistant in Medicine, Massachusetts General Hospital
Boston, Massachusetts

W. FREDERICK SAMPLE, M.D.
Chief, Section of Ultrasound and Computed Body Tomography
Radiology Department
University of California, Los Angeles
Los Angeles, California

ROLF L. SCHAPIRO, M.D.
Acting Head—Department of Radiology
The University of Iowa Hospitals and Clinics
Iowa City, Iowa

FRANK SEIDELMANN, D.O.
Associate Professor of Radiology
Case Western Reserve University
Cleveland, Ohio
Staff Radiologist, Marymount Hospital
Garfield Heights, Ohio

EDWARD V. STAAB, M.D.
Professor and Associate Chairman of Radiology
University of North Carolina
Chapel Hill, North Carolina

H. WILLIAM STRAUSS, M.D.
Director, Division of Nuclear Medicine
Associate Professor, Harvard Medical School
Associate Radiologist, Massachusetts General Hospital
Boston, Massachusetts

THOMAS A. VERDON, JR., M.D.
Director of Nuclear Medicine and Ultrasound
The Penrose Hospitals
Colorado Springs, Colorado

WAYNE W. WENZEL, M.D.
Director of Nuclear Medicine and Ultrasound
Presbyterian Medical Center
Denver, Colorado

***FRED WINSBERG, M.D.**
Diagnostic Radiologist in Chief
Montreal General Hospital
Associate Professor of Radiology
McGill University
Montreal, Canada

BRIAN W. WISTOW, M.D.
formerly: Assistant Professor of Radiology
Division of Nuclear Medicine
State University of New York
Upstate Medical Center
Syracuse, New York
currently: Assistant Professor, Department of Radiological Sciences
Division of Nuclear Medicine
University of California, Irvine Medical Center
Orange, California

VICTORIA S. YIU, M.D.
Diagnostic Radiology Resident
Department of Radiology
The University of Iowa Hospitals and Clinics
Iowa City, Iowa

Contents

Preface

The symposium on which this book is based was designed to answer many of the questions asked by physicians, third party payers, and the consumer as to the efficacy and proper utilization of the various imaging modalities. With the rapid growth of nuclear medicine, the explosive introduction of CT scanning, and the clinical acceptance of gray-scale ultrasound imaging, the majority of chapters on these modalities are presented in an enthusiastic and noncritical manner. In these papers, few studies demonstrate the true clinical indications for performing the various imaging procedures, and few compare critically the relative diagnostic weaknesses of these modalities. The need for this comparative data is obvious, and experts in diagnostic imaging are encouraged to establish the efficacy of the various modalities in clinical practice.

Since cost effectiveness is such an integral part of much of the new technology and its use today, the financial aspects of clinical diagnostic imaging are examined in detail. We believe these proceedings will be a landmark publication on the role of diagnostic imaging in the delivery of quality care to the patient.

We gratefully acknowledge and sincerely appreciate the interest and support rendered to the educational program presented in Anaheim, California, January, 1978, by the following suppliers and manufacturers of diagnostic imaging commodities.

The Harshaw Chemical Company, Solon, Ohio
Mallinckrodt, Inc., St. Louis, Missouri
New England Nuclear, Boston, Massachusetts
N.I.S.E., Inc., Cerritos, California
Picker Corporation, Cleveland, Ohio
E.R. Squibb & Sons, Inc., Princeton, New Jersey

We wish to thank the staff of the Nuclear Medicine Institute and the publisher for their time and diligence in the organization and publication of this book.

FINANCIAL CONSIDERATIONS

Commercial Aspects of Diagnostic Imaging

Ronald F. Meadows

Medical technology has progressed greatly in recent years and has made a major contribution to the improved quality of medical care in the United States, which is in itself, unquestionably, the finest in the world. New means of diagnosis and therapy have, to a considerable degree, transformed wide areas of medical practice. Continuous flow blood analyzers, open heart surgery, renal dialysis equipment, cardiac pacemakers, and, of course, computed tomographic (CT) scanning come immediately to mind.

At the same time, there has been increasing government concern about health care. Early on, the concern was with too little hospital capacity, which led to the federally supported Hill–Burton program in 1946 to expand it. Next came concern about limited access to the system and in 1966, Medicare and Medicaid came into being along with vastly expanded private insurance coverage. And now, government is concerned about the high cost of health care—which has come about in part because of excess hospital capactiy and almost unlimited access to the health care system. Obviously, costs must rise rapidly given such conditions—both excessive supply and demand.

The need to curb rising health care costs is widely expressed and few responsible people are in opposition to the call to "do something." What does provoke controversy, however, is the way in which health care costs can be most effectively and realistically brought to heel without risking doing damage to the quality of health care.

The development of computed tomographic (CT) scanning has closely paralleled recent rapid growth in health care costs, and thus has been cited by many as a key cause of rising medical costs. Largely prompted by CT's symbolism, medical technology has been cited as the engine driving hospital costs out of control, and CT scanners have been blasted as the new technological "toy" used by doctors and hospitals to rip off the public.

But, if CT is guilty of anything, it is the timing of its emergence—not its cost contribution. Rather than adding to health care expenditures, CT already shows real promise for containing and reducing medical costs—and since it has been in wide clinical use only several years, its potential for providing cost-effective, quality health care has barely been tapped. In fact, if it were not for the rapid development and acceptance of CT and the fact that it has become a symbol to many of medical technology gone rampant, it is unlikely in our opinion that proposed federal government caps on capital expenditures would have dealt with equipment capital expenditures at all. The facts indicate that medical technology and CT have been unfairly and erroneously singled out as major contributors to rising health care costs.

Overall health care costs rose approximately $80 billion, or an average of 11.8 percent per year between 1965 and 1975 (Fig. 1). In 1976, costs grew over 15

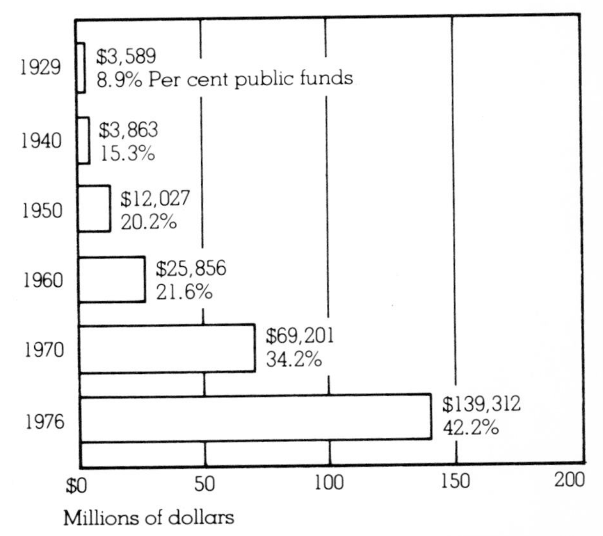

Figure 1. Expenditures for health care. (Sources: Social Security Administration and Tax Foundation, Inc. by permission of Technicare Corporation Annual Report, 1977.)

percent. Community hospitals stand out by representing $30 billion of the $80 billion growth, or 38 percent. Since the criticism of technology costs seems to focus on hospitals, the discussion will, from this point on, deal with the makeup of that $30 billion.

Components of Cost Increase

Hospitals, of course, have not been spared from inflation (Fig. 2). Of the $30 billion increase in the costs of community hospitals, 45 percent, or $13.5 billion, was due to inflation. Yet, there is not a great variance between the growth of the hospital price index and the consumer price index (Fig. 3). The hospital price index went up 6.7 percent annually in the ten-year period, while the consumer price index rose 6.1 percent.

The second most important component of the $30 billion is increased usage or patient days. This accounted for 15 percent, or $4.5 billion, of the $30 billion increase. In the ten-year span, admissions in hospitals went up 7 million and outpatient visits increased over 100 million; weighting the two, we see a 30 percent increase in overall hospital admissions. And, of course, more admissions translates into more beds.

Medicare and Medicaid are major reasons for the increase in hospital patients and costs. For instance, for the poor in the 65-and-over age group, the number of hospital trips per hundred people went from 18 to 25; in the not poor, over-65 age group, the rate went from 20 to only 23. This represents a major increase in the services hospitals have been asked to provide—and it represents added cost.

Therefore, of the $30 billion increase that we are attempting to analyze, $12 billion remains to be explained. This unexplained amount is often labeled as medical inflation, the average cost per patient day. It is an exceedingly difficult figure to understand. Yet, it is the inability to understand the $12 billion that seems to give rise to most of the charges against medical technology.

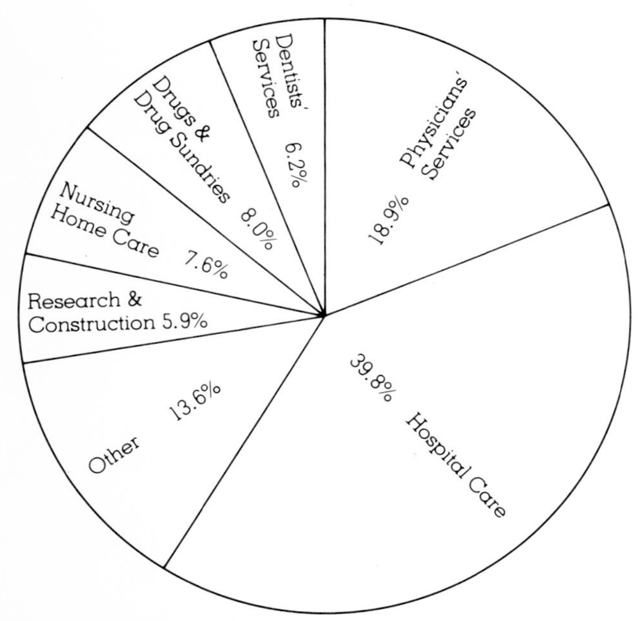

Figure 2. National health care expenditures by type of expenditure–1976.(Source: Department of Health, Education, and Welfare, Social Security Administration by permission of Technicare Corporation Annual Report, 1977.)

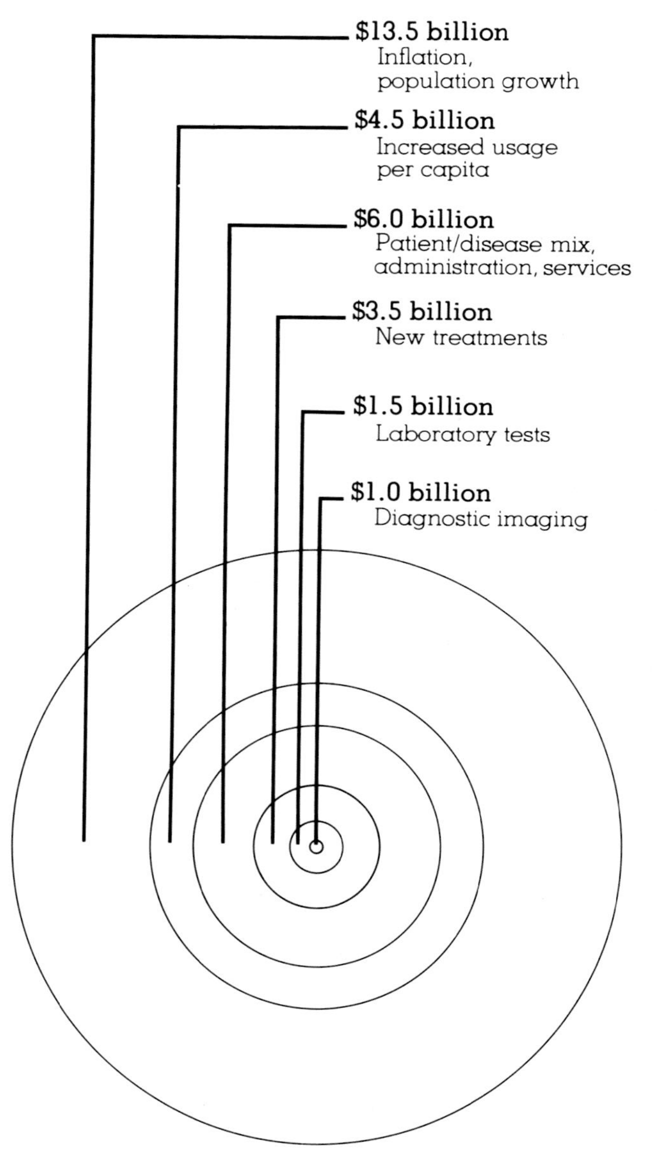

Figure 3. Components of $30 billion hospital cost increase, 1965–1975 by permission of Technicare Corporation Annual Report, 1977.

Table 1. Cost increase for treating aged*

Age group	Percentage of 1975 population	Percentage of hospital expenditures		Per capita hospital cost ($)	
		1967	1975	1967	1975
Under 19	35	10	11	22	71
19 to 64	55	65	60	104	230
65 and over	10	25	29	224	603

* Source: Social Security Administration.

It appears (after considerable fact finding and analysis) that the $12 billion in intensified medical inputs breaks into two roughly equal parts, about $5.5 to $6.5 billion each for technology and what we have labeled "other inputs." The "other inputs" fall into three general categories: (1) more complex patient/disease mix; (2) administration and overhead; and (3) more intensive nontechnology inputs.

As far as patient mix is concerned, the shift to the treatment of the aged via Medicare shows up vividly (Table 1). In 1975, the 65-and-over age group represented only 10 percent of the population but 29 percent of hospital admissions. Why? Because it's expensive to treat the aged. In 1975, the cost of treating the 65-and-over group was $603 per patient, up from $224 eight years previously and three times the per capita cost of the 19 to 64-year-old segment of the population.

Disease also has become more complex and costly to treat. Obviously, there has been important progress in the treatment of disease, such as diseases of early infancy and tuberculosis. But some of the diseases that have not yet been conquered pose major hurdles. Heart disease, cancer, and stroke account for over 600 of every 1000 hospital deaths and, in many ways, they are the most expensive and difficult to diagnose and treat.

The second factor to consider is administration and overhead. Dr. John H. Knowles, President of the Rockefeller Foundation (and a Director of Technicare) authored an article for *Scientific American* analyzing the cost patterns at Massachusetts General Hospital between 1964 and 1972, during which time he served as Chief Administrator of the Hospital. In those eight years, administrative costs at that hospital increased seven times, or 60 percent faster than patient-related expenses.

Examples of administrative overhead are: larger debt and depreciation—in 1963, hospitals financed only 21 percent of their construction with debt versus 50 percent today; expanded data processing; and more regulation—in New York State alone, 99 regulatory agencies govern hospital operations.

The third factor, more intensive nontechnology inputs, includes expanded services such as social work, physical therapy, and abortion services; the increase in number of employees per patient, adjusted for outpatients, from 2.25 ten years ago to 3 today; and, the increased use of drugs, supplies, and disposables. Again, the Massachusetts General experience is relevant: nontechnology input costs went up 300 percent in the eight years between 1964 and 1972, growing at a 14.5 percent compound annual rate, substantially faster than increases for overall health care.

The technology half of the $12 billion cost increase from intensified inputs is comprised of three categories—treatment, diagnostics, and routine lab testing. Keep in mind that this $5.5 billion figure represents only 7 to 8 percent of the $80 billion increase in health care costs in the last ten years (Table 2). Over half of the technology component, $3 to $4 billion, reflects new methods of treatment. And it is these new techniques, such as organ transplants, intensive care, renal dialysis, implants, burn centers, and radiation therapy, that unquestionably extend life.

The balance of the $2.5 billion technology component is comprised of $1 billion in diagnostic imaging

Table 2. Cost increase for new life extending treatments, 1965–1975*

	Estimated cost increase ($ billions)
Surgical procedures (open-heart, organ transplant)	1.0–1.5
Life support—short term (intensive care)	0.5–1.0
Life support—long term (renal dialysis)	0.5–0.5
Life improving surgery (implants)	0.3–0.6
Emergency (burn centers)	0.2–0.5
Therapy (cobalt treatments)	0.1–0.3
	2.6–4.4

* Source: Booz, Allen & Hamilton estimates.

Table 3. Cost increase for diagnostic imaging, 1965–1975*

		$ Millions
More use		400
Expansion to more hospitals		
Conventional x-ray	215	
Nuclear medicine	235	
		450
New technology		
Ultrasound	100	
CT scanning	50	
		150
		1,000

* Source: Booz, Allen & Hamilton estimates.

and $1.5 billion in laboratory testing. Each of these includes more use of older methods as well as new technology.

In the 1965–75 period, radiology expenses went up a total of $1.7 billion. But again, inflation and increased admissions accounted for $0.7 of the $1.7 billion increase. Thus, we see a net $1 billion increase in radiology expense due to technology and more use.

The radiology cost increase is comprised of three components. First, more use, i. e., more procedures per admission (Table 3). This amounts to $400 million. The second factor is expansion to more hospitals of technologies in use prior to 1965, particularly conventional x-ray and nuclear medicine. This amounts to $450 million. The third component is the use of CT scanning as seen below.

Cost of CT Scanning

The really new diagnostic imaging procedures—ultrasound and CT scanning— account for the remaining $150 million. Obviously, the time frame of 1965–75 does not accurately portray the cost of CT scanning, since it was introduced in 1973. By 1975, only about 100 CT scanners were operational in the U.S., representing about $50 million in patient charges.

The year 1976 provided a more realistic assessment. It appears that patient charges for CT scanning in 1976 increased to about $200 million. There were about 300–350 scanners operational, running close to practical capacity of about 2000 patients per scanner, and the average charge per scan was about $240, including the physician cost. This adds up to patient charges of somewhere between $180 and $210 million, or about $200 million rounded off. So while CT costs increased fourfold, they were less than 1 percent of the $20 billion health care cost increase in 1976.

Looking ahead to 1980, we estimate that patient charges for CT scanning in 1980 will be in the $750 million to $1 billion range, or less than half a percent of total health care costs. This is based on a projection of 2000 scanners in place or about one per 110,000 people. We also estimate patient throughput at about 2000 examinations per year.

As is well known by the medical profession, there are practical limits in hospitals that prevent substantial increases in CT utilization. First, help is difficult to come by at night. More importantly, inpatients—particularly the ill—cannot freely be taken to radiology in the middle of the night for a scan. And outpatients generally would not come in at odd hours. Finally, body scans typically require 75 minutes versus only 50 minutes for head scans, and there are more and more body scans being done.

Cost/Benefit of CT Scanning

We have dealt thus far with gross CT costs rather than net costs. But the key point, which largely has been lost on the critics of CT scanning, is that in terms of productivity or cost/benefit, CT has enormous potential to contain and reduce health care costs. CT scanning provides physicians with an effective means to help achieve a goal we would presume just about everyone would endorse—"straight line" medicine. That is, a physician should attempt to diagnose a disease or abnormality quickly, precisely, and at lowest cost to the patient and should eliminate more costly and less revealing traditional procedures whenever possible. Thereby, a sound program of patient care can be prescribed sooner.

CT scanners can help make significant reductions in diagnostic costs in two ways. Over the long term, as physicians become more experienced with its capabilities, it will be used increasingly as a procedure of first choice. Therefore, there has been and will be increasingly a reduction or outright elimination of some other more costly and risky techniques, including exploratory surgery. Second, CT scans can be performed on outpatients, while cost comparable procedures require hospitalization.

A study just completed by Arthur D. Little, Inc., a research and consulting firm, and Dr. Ronald Evens, Director of the Mallinckrodt Institute of Radiology, St. Louis, projects that if CT scanning had not existed in 1977, a total of 9,120,000 diagnostic procedures would have been made (in the head and areas of the body relevant to CT's potential capabilities), at a cost to society of slightly over $3 billion. However, CT did exist and 1,500,000 CT procedures were in fact conducted. But actual total diagnostic costs were only slightly more than the pre-CT figures.

In 1980, with wider availability of CT, 4 million CT procedures (267 percent increase) are estimated (for the same population base and in 1977 dollars), and total diagnostic costs will be less than in 1977. The conclusion: The important medical benefits of CT are attainable with no apparent increase in health care costs.

Regulatory Controls

As a result of legislation already established, high technology medical equipment that sells for more than $100,000, such as CT scanners, is one of the most tightly regulated sectors of the health care industry. At the present time, every state except Missouri has some statewide regulatory authority over hospital purchase of CT scanners and nearly four-fifths of the states have CT planning guidelines in effect or under development. At last count, 29 states and the District of Columbia have Certificate of Need laws and 20 review capital expenditures under 1122. The combination of the 1122 and CON programs makes it virtually impossible for a hospital to purchase a CT scanner without authorization.

In fact, Health Service Agencies are becoming very effective in limiting CT placements, as partly evidenced by the reduction of orders for new scanners from about 600 in 1976 to about 240 in 1977. Based on our surveys of HSAs, we also estimate that the vast majority of CT scanner Certificate of Need proposals have been turned down in the last 12 months. These figures differ from some statements I've seen that HSAs are relatively ineffective and, specifically, that they "approve nine out of ten CT applications."

But, HSAs need more information on which to base sound recommendations—such as CT efficacy, cost effectiveness, "common denominator" data on present equipment utilization and cost effectiveness, and the experience of other HSAs.

Numerous legislative proposals have been advanced to limit growth in health care costs. They shall not be described here in detail; suffice it to say that many people are busily engaged in trying to legislate controls on hospital costs.

However, we at Technicare are convinced that cost-conscious hospital management, in combination with existing planning and regulatory procedures, can be effective in controlling health care costs.

We also believe that incentives to achieve hospital cost consciousness, such as Senator Talmadge proposes, represent a constructive approach and deserve strong public support. They would reward good hospital management and penalize the less efficient and help assure a continuation of quality care.

But the proposals we have seen for controlling capital spending by hospitals are geared neither to cost efficiency nor to quality of care, and in many instances will be in direct conflict with the aim of restraining operating costs. The facts suggest that for whatever the reasons, the nation is faced with excess and inadequately distributed hospital bed capacity. However, some of the steps proposed to do something about bed capacity lump this problem together with medical equipment capital investment, as if these were the equivalent of bricks and mortar. All capital spending is condemned out of hand. This also says that no attempt is made to differentiate between equipment that enhances the quality of care and improves efficiency, and equipment that would simply add to operating costs.

Medical equipment capital investment—products ranging from typewriters to radiation therapy equipment—appears to represent less than 20 percent of total medical capital spending. Consequently, Congress runs the risk of seriously jeopardizing the quality of health care with minimum potential reward in terms of reduced capital expenditures.

The CT scanner is now meeting both the test of medical efficacy and the test of cost effectiveness. We believe that CT should be judged on its ability to meet these tests, not arbitrarily condemned due to misunderstanding concerning its contribution to diagnostic medicine, costs, and utilization.

Right now, regulators seem positioned for overkill because constraints on CT scanning purchases are already substantial. Congress ought to be very cautious not to pick the wrong target because of its popularity, to balance the quality of care equally with cost consciousness, and to make economic judgments based on true costs.

Climate for Efficiency

The regulatory climate in health care in this country today is one that I think none of us would have even imagined a couple of years ago. There has also been a new awareness of the need to be more cost conscious. It has caused hospital administrators and physicians, I believe, to be more cost conscious, particularly with respect to how capital is invested. And manufacturers see a greater sophistication and awareness on the part of doctors and hospitals of what the true cost of an investment is—factors beyond the initial purchase price. I refer to such elements as patient throughput, cost effectiveness of a particular piece of equipment, reliability, etc.

From our point of view as a manufacturer, we are trying to respond to cost containment pressures, as well. Most obviously, we respond through product design. We try to design equipment that embodies greater productivity-to-cost ratios. For example, in nuclear medicine, our Video Image Processor, introduced just several months ago for cardiac diagnosis, is available at $40,000 versus a similar competitive product which costs $60,000 to $80,000.

Ohio-Nuclear's new Real-Time Compound B Scanner, Sonofluoroscope One, sells for about $30,000, half of the cost of previous products. Of course, our new neurologic CT scanner, which begins as low as $96,500, is an obvious response to cost pressures.

Incidentally, we've been asked if it is possible to reduce the costs of body scanners to a similar degree. I think we have to say in all honesty that it isn't possible because of the enormously greater amount of sophistication associated not only with the need for faster scanning speed, but with the capability of processing the data to provide the kind of image necessary in as short a time as possible. This isn't to say that body scanner prices cannot come down somewhat, but not to the degree of reduction seen in head scanning equipment.

Another way we try to respond, of course, is to make our salesmen more capable of dealing with issues of equipment productivity related to hospital capital investment problems. We spend much more time teaching them some of the rudiments of finance and economics.

This doesn't mean that buyers are not still entitled to put pressure on us for a lower price in a given situation. It also doesn't mean that we are still not entitled to say no.

In any case, there seems to be a climate for more cost consciousness in health care which is beneficial for us as manufacturers, for doctors, and for patients as well.

Financial Aspects of Ultrasound

Gordon S. Perlmutter

To date, there is very little information in the literature concerning the financial aspects of operating a diagnostic ultrasound laboratory. A recent publication by the Alliance for Engineering in Medicine and Biology,[1] however, does outline some of the finances of establishing and maintaining a diagnostic ultrasound laboratory. These figures, however, vary considerably with the author's experience. Because of this, an informal survey of several community hospitals of varying size as well as a few academic centers in eastern Pennsylvania was performed.

The figures to be presented in this section represent average figures developed from this informal survey. Preliminary requirements for the operation of an ultrasound laboratory are summarized in Table 1. An average acute care community hospital with an average ratio of inpatient to outpatient studies of 1 : 1 can anticipate a caseload of approximately two cases per one hundred hospital beds once the diagnostic ultrasound laboratory has become well established. A period of about 6 to 12 months can be anticipated in most institutions from commencement to full utilization of ultrasound services, depending upon how vigorously the ultrasound services are promoted to the general medical staff. Anticipated equipment requirements would include one fully equipped gray-scale B mode scanner per 8 to 10 patients examined. This is an average figure, however, and some centers report performing as many as 14 examinations per scanner per day during peak periods. Personnel and space requirements are given in Table 1.

Table 1. Preliminary requirements

Volume anticipated	2 cases/100 beds
Equipment	1 scanner/8 to 10 patients
Personnel	1 technician/8 to 10 patients
	1 secretary-receptionist
	150 ft^2/scanner
Space	300 ft^2/support facilities

An outline of capital equipment costs for establishing an ultrasound laboratory is given in Table 2. Comparison of conventional and digital gray-scale scanners has been made elsewhere in this text in the chapter on Advances in Ultrasound. Although not listed in Table 1, it becomes increasingly apparent that state-of-the-art real-time ultrasound units are becoming a required addition for any modern ultrasound laboratory. Most of the hospitals surveyed have one or more of the less expensive real-time units listed in Table 2. Because of its expense and limited field of view, the phased array unit listed in Table 2 has received only limited application for real-time cardiac imaging. One can anticipate in the near future, however, that sophisticated real-time units in the same price range as the phased array unit will appear on the market with resolution characteristics equivalent to currently available gray-scale scanners. Multiformat image cameras using radiographic sheet film are a must for any ultrasound laboratory, in that capital costs of the multiformat camera and the x-ray film processor are rapidly offset by the high cost of alternatively using self-developing film.

Projected revenues and expenses for the operation of a hypothetical 400-bed acute care facility are given in Table 3. Gross revenues are predicated on a caseload of 8 patients per day or 2080 patients per year, given an operational year of 260 days. An average charge per scan of $80 has been used for predicting gross reve-

Table 2. Equipment costs($)

Gray scale scanner	
Conventional	44,000
Digital	55,000
Cardiac unit	20,000
Real-time	
Sector scanner	12,000
Linear array	22,000
Focused array	32,000
Phased array	80,000
Multiformat camera	6,000
Processor	4,000

Table 3. Projected revenues and expenses($)

Basis: 400-bed acute care facility	
Gross revenues (@ \$80 × 2080)	166,400
Deductions for free care/bad debts	21,600
	144,800
Expenses	
Personnel (1 technician/1 secretary)	23,000
Depreciation (\$87,000/5 yr)	17,400
Maintenance	3,500
Supplies (add \$11,400 for Polaroid)	4,700
Indirect expenses	5,800
	54,400
Net income	90,400

Table 5. Projected revenues and expenses for additional cardiac unit($)

Basis: 400-bed acute care facility	
Gross revenues (@ \$80 × 520)	41,600
Deductions for free care/bad debts	5,400
	36,200
Expenses	
Personnel (½ technician)	7,000
Depreciation (\$20,000/5 yr)	4,000
Maintenance	1,000
Supplies	1,500
Indirect expenses	1,600
	15,100
Net income	21,100

nues, however, this figure was found to be highly variable, depending upon the instituions surveyed and the ratio of the various types of scans performed. For example, the average cost per scan tended to be lower in hospitals having a high percentage of obstetrical cases compared to those performing relatively less obstetrical studies. Depreciation is predicated on straight line depreciation of \$87,000 capital equipment over a five-year interval. Note should be made of the rather substantial increase in the supply costs when self-developing film is used as an alternative to x-ray film in a multiformat camera. The capital expenditure for the multiformat camera has been included in the \$87,000 basis for depreciation. The remaining items in Table 3 are self-explanatory.

Figures for the operation of a cardiac ultrasound laboratory have been segregated from the operating figures of a general ultrasound laboratory and are given in Table 4. Two reasons for doing this are that, in several institutions, the cardiac ultrasound laboratory is run as a separate entity, usually by the cardiology service. However, even when cardiac ultrasound is offered as part of a general ultrasound laboratory service, usually under the aegis of the Department of Radiology, it tends to operate as a segregated section in terms of patient load, equipment, and personnel. The anticipated volume of patients for a cardiac unit is considerably smaller than that for a general diagnostic unit as

Table 4. Additional requirements for cardiac unit

Volume anticipated	1 case/200 beds
Equipment	1 unit/4 patients
Personnel	1 technician/4 patients
Space	150 ft²/unit

indicated by the anticipated one case per 200 hospital beds as listed in Table 4. Specialized cardiac ultrasound laboratories do exist that perform severalfold more examinations per hospital bed than are listed in Table 4, however, no such institution was included in the survey that served as the basis for the figures given in this section.

Returning to the hypothetical 400-bed acute care facility used as the basis for the figures in Table 3, data have been projected for the operation of an additional cardiac unit and are given in Table 5. Gross revenues are based upon an average cost per scan of \$80 and an average patient load of 2 patients per day or 520 patients per year. The remaining items in Table 5 should be self-explanatory.

Note should be made that both in Tables 3 and 5, the additional expense of performing real-time ultrasound scanning was not included in the depreciation figures. Of the institutions surveyed, too few were using real-time units to accurately assess the financial impact of this equipment on the operating budget of an ultrasound laboratory. Those institutions using real-time units generally offered a real-time study as the only examination in only specific instances, such as for portable ultrasound studies in the delivery room or for bedside evaluation of the patient with suspected aortic aneurysms. In these instances, the charges for the real-time studies were generally less than those for comparable gray-scale ultrasound evaluation. In all other instances, real-time studies, performed in conjunction with routine gray-scale scans, were done at no additional cost to the patient.

It should also be noted that the average cost per scan listed under gross revenues in Tables 3 and 5 represents total components, with no attempt to break down charges on the basis of technical and professional charges. Also, because of the wide variation in billing practices in the hospitals surveyed, no attempt was made to list professional fees as an expense item. Professional fees, therefore, must be derived from the net income figures given in Tables 3 and 5.

Figures derived from the institutions surveyed that satisfied the prerequisites mentioned above do not vary significantly from the averages given in the tables of this chapter. It should be noted, however, that specialty institutions, institutions with a wide disparity between inpatient and outpatient volume, and chronic care facilties had figures at wide variance with those given in the tables.

These statistics are offered as a guideline for planning the financial aspects of a diagnostic ultrasound facility. They are in no way intended to indicate what fees can or should be charged for ultrasound services. Patient charges must be developed on an individual institutional basis after careful analysis of the operating expenses of the ultrasound laboratory applicable to that facility.

Reference

1. System Design of a Clinical Facility for Diagnostic Ultrasound. Bethesda, Md., The Alliance for Engineering in Medicine and Biology, 1977

Financial Aspects of Computerized Tomography

Gerald S. Freedman

The introduction of computerized tomography has had a profound effect on the practice of diagnostic radiology and the health care industry.[3,6,11] Born of high technology at a time when the U.S. health sector of the economy was feeling the first constraints on unbridled growth, it has been singled out as the one item that represents all that is good and all that is potentially bad in the U.S. health care system.

CT demand escalated as the profession became aware of its diagnostic potential and the educated consumer developed increasing diagnostic expectations. Manufacturers of high technology, medical equipment or other manufacturers in related fields with capital for research and development sought to cash-in on this new item. Unfortunately, all of this happened at a time when federal and local health planning agencies were seeking to contain the everincreasing cost for health care. In this chapter we will seek to evaluate those financial forces and aspects of the health care system that are simultaneously stimulating expensive technology (consumer and industrial input) and constraining expenditure as applied by government, state, and to a lesser extent the insurance industry.

Historical Background

If we look back on the growth of the health care industry over the last 50 years, we find about a 50-fold increase in cost (Table 1). According to the Social Security Administration, $3,589,000,000 were expended in the U.S. in 1929. Of this, 8.9 percent were public funds. The same source states that in 1976, $139,312,000,000 were spent, and of these 42.2 percent were public funds.[1]

Of this amount spent in 1976, 40 percent went to hospital care, 19 percent to physician services, 8 percent to drugs, 6 percent for research and construction, 6 percent for dental services, 7.5 percent to nursing homes, and 13 percent elsewhere. This type of escalation in health care costs and increasing public reimbursement has given rise to ever greater Federal involvement.

Table 1. Survey of health care costs in US since 1929

Year	*Millions of dollars*	*Percent public funds*
1929	3,589	8.9
1940	3,863	15.3
1950	12,027	20.2
1960	25,856	21.6
1970	69,201	34.2
1976	139,312	42.2

Federal Health Planning History

In 1964 the Hill–Burton program was passed which made funds available to build hospitals, thereby increasing potential medical services. This was followed by the Comprehensive Health Planning Act of 1966 which resulted in local health planning councils and initiated the Certificate of Need review process. In 1972 a section of the Social Security Act authorized individual states to control capital expenditures in health services. Most recently, the National Health Planning and Resources Development Act in 1974 required all states to develop Certificate of Need laws which would be compulsory in the year 1980. Currently, there are numerous legislative proposals before the 95th Congress. Under consideration are the Cost Containment Act of 1977, introduced by President Carter, and the Medicare–Medicaid Administrative and Reimbursement Reform Act, introduced by Senator Talmadge. Proposals by Senator Schweiker of Pennsylvania and others are also under consideration.

Title One of the Carter proposal is intended to be a

temporary hospital cost containment program to limit the increase in hospital revenues to approximately 9 percent. Title Two of this proposal would impose a national ceiling of $2.5 billion a year on capital expenditures over $100,000 by hospitals. As a point of reference, the $2.5 billion represents about one-third of the estimated 1977 capital expenditures in the hospital industry. If we assume $0.5 million per scanner, then 500 scanners would cost $250 million, or 1/10 the potential approved expenditure.

The Talmadge proposal deals more with methods of reimbursement for hospital operating costs under the Medicare and Medicaid Program. Its intent is to reward those hospitals whose costs are below average and to penalize those whose costs are high. This approach leaves the question of capital investment to the hospital, to decide what course to choose to maximize efficiency. There seems to be much good sense in this approach since the current reimbursement does not foster innovative, efficient health care.

Another bill worthy of mention is the Kennedy proposal introduced in the Senate, which suggests an immediate moratorium on all individual hospital capital expenditures over $150,000 in those states without HEW approved health planning processes. At the moment if this bill were to be passed, it would result in a national de facto moratorium since HEW has not issued appropriate guidelines for the states to follow and, therefore, no state plans have yet been approved.

It is my view, coming from a state (Connecticut) that has particularly stringent Certificate of Need laws with presumed jurisdictions over and into the private sector, that stringent control is inevitable.[9] The determination in Connecticut as to whether or not the state in fact has their presumed jurisdiction over the private sector is being tested in the courts.* Even if the courts find in favor of the private sector, this is likely to be a hollow victory. An alternate, more secure method of captial expenditure control can be achieved through reimbursement control. It is likely that certain insurers, such as Blue Cross[2] and Blue Shield, will seek to limit their cost exposure by reimbursement only to those facilities that have acquired a Certificate of Need (CON). Furthermore, if the case is won by the private sector in the state court, I suspect that the legislature will rewrite the law to cover the private sector.[12]

Current CT Installations And Projections

The forces described on the preceding pages have abruptly halted the rapidly expanding CT market. Data from J. Lloyd Johnson[4] Associates show that it took two years following its introduction in 1972 before the first 100 units were ordered. At the end of 1975 there were only about 200 units installed in the United States. By mid-1976 installed units numbered 320, with an additional 200 already approved for installation. By mid-1977 over 600 units had been installed and another 250 had been approved. The best estimate as of January 1, 1978 is a total of about 1000 units installed and approved and another 1000 applications pending.[8]

The U.S. market in 1975, as prepared by Kidder, Peabody and Company, was estimated at a total of 500 units installed by the end of 1976, toward a total of 1425 units in place by 1980. The growth curve was well on the way to achieving that until it encountered Federal and State Certificate of Need laws imposed during 1976. Whereas, in 1976 over 400 orders were taken by 15 companies, the estimate for new orders taken in 1977 is believed to be closer to 200 at this time. In 1975 Kidder, Peabody and Company anticipated that the top five companies were to have captured 90 percent of the scanner market by 1980 in the following fashion: EMI, 26 percent; Technicare, 25 percent; Pfizer, 20 percent; General Electric, 18 percent; Picker, 4 percent; and others, 7 percent. Certainly in the past year, Technicare and GE have further improved their units and market potential. Although EMI has continued to improve their products, they no longer enjoy the preeminence that they initially had. It appears unlikely to me that Pfizer can maintain a relative parity position to the other three, and clearly Kidder–Peabody's expectations of the Picker unit were not forthcoming. Furthermore, it would appear that Technicare, EMI, and General Electric have each probably developed almost sufficient in-house manufacturing capability to meet the needs of the entire U.S. market. Also, the remaining market outside the U.S. is smaller and even more competitive. The Kidder–Peabody's report estimated the entire foreign market to be about equal to the U.S. market, with 90 percent of those sales occurring in Western Europe or Japan. Clearly, something major is about to take place in the CT market!

Table 2 lists 15 companies whose advertisements indicate they are in the CT market. The equipment available is listed by the commonly listed generation nomenclature, 1st, 2nd, 3rd, and 4th; T/R is Translate-Rotate. In addition to the ones already mentioned, there are Syntex, Varian, Artronix, Philips, El Scint, A.S.E., Searle, C.G.R., Siemens, and Hitachi. If certain foreign companies such as Hitachi, C.G.R., Siemens, and Philips can, in fact, capture substantial portions of the sales in Japan and Europe, then the companies that might have the greatest marketing difficulty are; Syntex,* Varian,†

*December 22, 1977 lower court ruling in favor of private physicians' right to purchase a CT scanner appeal expected.

*Philips has entered into an agreement to purchase Syntex head units for sale outside the United States.

†C.G.R. has entered into an agreement to market Varian scanners.

Table 2. Companies currently in CT market and units available*

Company	1st (T/R slow)	2nd (T/R fast)	3rd (rotate)	4th (stationary)
Technicare	X	X		X
EMI	X	X		X
Pfizer	X	X		(X)
GE	X		X	
Syntex	X	X		
Varian			X	
Artronix			X	(X)
Phillips		X	(X)	
El Scint		X		
A.S.E.				X
Picker		X	X	
Searle			X	
C.G.R.		X		
Siemens	X		X	
Hitachi	X			

* X, available; (X), under development.

Table 3. Cost data

	Body CT location		
	Hospital	Outpatient	Average all units
Equipment cost ($)	530,000	496,000	526,000
Years depreciated	5.7	5.8	5.7
Accelerated depreciation	0/47	4/5	4/52
Space allocation (ft^2)	653	1078	694
Remodeling cost ($)	58,000	37,500	55,000
Space cost ($/$ft^2$/yr)	12.18	18.15	13.52
Technologists	1.9	2.0	1.9
Other personnel	1.0	1.1	1.0
Contrast cost ($)	6.95	6.04	6.83
Other supply cost ($)	12.70	12.36	12.66

A.S.E.,* Picker, Artronix, and Searle; although several of these units produce excellent images.

Economic Analysis of a CT Scanner Installation

Having evaluated the various companies involved in the CT market, we must now begin an analysis to establish the relationship of cost of a piece of equipment, other direct and variable costs associated in running a facility, and the indirect costs related to the operation. Tables 3 and 4 compare two surveys, one by Evens[7] and the other by Health Service and Management. Both list cost components for the annual operations of a CT unit. The Evens head equipment is representative of a low cost unit, and the Health Service Management Unit is of intermediate cost. The biggest difference is the applied indirect cost. Evens assumed indirect costs to be 50 percent of the total direct costs. This is for a large teaching hospital. Health Service Management has an indirect cost of less than one-third of the direct cost.

Based on these data, Table 5 attempts to draw a more comprehensive economic analysis to be used with scanners of every cost.[10] From Tables 3 and 4 we were able to justify an indirect cost application of $50,000 independent of the facility. A second category of essentially fixed costs, including personnel, supplies, space, and remodeling, approximates $100,000. If we then consider the spectrum of equipment costs from the low end* of $225,000 to the high end of $700,000, we can determine a total direct and indirect cost. The difference is about a factor of two. At the lowest end we are looking at an annual cost in the neighborhood of $200,000 and at the highest end, something over $400,000.

*A law suit claiming a priority to stationary detector technology based on certain patent claims, if granted, could substantially improve the outlook for A.S.E.

*At the 1977 RSNA a head CT unit was introduced at $96,000.

Table 4. Comparative operational cost estimates for CT scanners ($)

Item	*Evens*[4] *(head)*	*Evens*[7] *(body)*	*Health Services*[4] *Management (head)*
Direct cost			
Fixed			
Equipment	77,400	105,200	117,000
Remodeling	3,860	11,000	10,000
Maintenance contract	25,000	30,000	20,000
Personnel	49,200	36,500	58,000
New development	25,000	25,000	10,000
Space upkeep	6,229	9,383	—
	186,689	217,083	215,000
Variable (supplies)	37,960*	44,460	37,300†
Total direct costs	224,649	261,543	252,300
Indirect costs	112,324	130,822	37,500
Total costs	336,973	392,365	289,800

* Based on 50 patients per week.
† Based on average of 56 patients per week.

The graph in Figure 1 plots all these variables together. A break-even analysis can be easily determined based on the number of patients seen per day, the charge per study, and the equipment used. For example, equipment that cost $450,000 has an associated $300,000 annual total operating cost. If one charges $200 a study (and collects), excluding physician's fees, one would require an average of about 6.5 patients per day to achieve the break-even point. On the other hand, if $150 is charged (and collected) per study, then 9 studies per day would be needed. Let me further underscore the point that this break-even analysis *assumes* 100 percent payment. Because of the complex and widely variable payment schemes, 80 percent reimbursement is more realistic. Therefore, one would require approximately 8 studies per day at $200 per study

Table 5. Representative economic analysis of CT scanners

Direct costs ($)

Equipment cost	Annual cost*	Update & maintain	Equipment total/yr	Total Direct & indirect
225,000	61,000	18,000	80,000	230,000
350,000	96,000	28,000	124,000	275,000
450,000	122,000	36,000	158,000	300,000
550,000	155,000	44,000	199,000	350,000
700,000	192,000	56,000	250,000	400,000

Other fixed expenses:

Personnel (secretary, technician)	50,000
Supplies (10 cases per day)	35,000
Space (rental)	10,000
Remodeling	5,000
	100,000

Indirect costs ($)

Administration, utility, insurance	50,000

* Acquisition cost (lease 5 years, 2.3% per month).

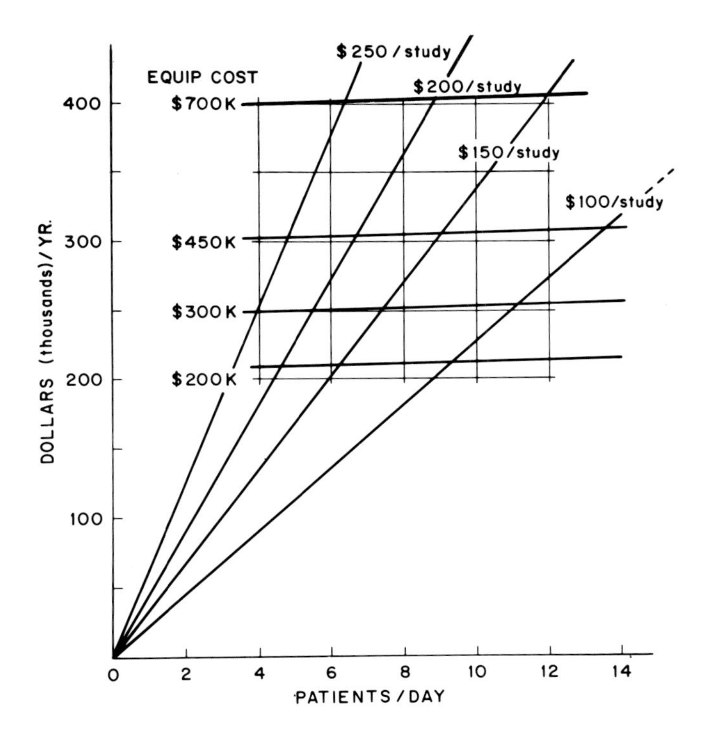

Figure 1. Break-even CT scanner analysis using the following variables: initial equipment cost, annual operating cost, number of patients per day, charge of study. (Reprinted from Applied Radiology, Vol. 7, No. 6, November/December 1978, copyright by Barrington Publications, Inc., 825 S. Barrington Ave., Los Angeles, California 90049).

or approximately 11 studies per day at $150 per study to attain a break-even operation.

If we turn to Table 6, we see a similar economic analysis of a $450,000 unit based on $200 per procedure. It is determined on the basis that income is 80 percent of billing and considers 5 different patient loads per day. It would appear that with 10 patients per day and operating 250 days per year, a unit could show potential profit of about $100,000 per year excluding professional fees. Note that at 6 patients per day a $40,000 annual loss could be expected.

The whole question of CT reimbursement is a very complex one. Table 7 attempts to summarize the various components. For this purpose it is important to divide the problem into head scans and body scans as well as units located in a hospital setting and units located in a private setting. If we begin by looking at the Blue Cross arrangement, they pay charges for head and body work in a hospital setting. This is true for all patients having studies performed in a hospital! Inpatients are reimbursed as part of the agreed per diem for each hospital as negotiated with each hospital. *Blue Cross does not pay for any CT study performed in a private setting, regardless of whether that private setting has or has not a Certificate of Need!* Blue Shield is an extremely complex situation and the reader is encouraged to contact his Blue Shield representative in his state for the exact details. In Connecticut there is no payment by

Table 6. Economic analysis of a CT scanner*

	1st	*2nd*	*3rd*	*4th*	*5th*
Number of procedures per day	6	8	10	12	14
Cases in 250 operating days	1,500	2,000	2,500	3,000	3,500
Billing ($)	300,000	400,000	500,000	600,000	700,000
Income (80% of billing) ($)	240,000	320,000	400,000	480,000	560,000
Expenses ($)	280,000	290,000	300,000	330,000	350,000

* Acquisition cost = $450,000, charge per procedure = $200.

Table 7. Total CT Reimbursement with contrast

	Head		*Body*	
	Hospital	*Private*	*Hospital*	*Private*
Blue Cross	Charges	No	Charges	No
Medicare	$172.50	$172.50	$193.50	$193.50
Medicaid	Per diem	Amount not determined	Per diem	$150.00*
Blue Shield	Under evaluation in Connecticut			
Private insurance	Yes—variable			
Institute of Medicine[2]	(Cost based on 5 years at 2500 patients/year)			

* Connecticut.

Table 8. Typical CT charges ($)

Study	*Facility*	*M.D.*	*Total*
Head*			
No contrast	157	63	220
Contrast	187	74	261
Body†	222	70	292

* No contrast head studies averaged 9 scans in 32 minutes; contrast studies averaged 12 scans in 53 minutes. From R. Evens, RSNA 1977.

† Body scans averaged 11 scans in 50 minutes; with contrast the studies required 15 scans in 77 minutes. From the American Hospital Association 1977.

Blue Shield (CMS) at this time (December 1977). Table 8 lists typical charges.

It is reported that Blue Shield has proposed to the State Insurance Commission a fee schedule including payment for CT work done in private facilities that have a Certificate of Need. The amount of reimbursement is not known at this time, but includes coverage for outpatients and hospitalized patients transferred to the private facility. Medicare will pay for head images only. They currently have a recommended maximum professional fee of $55 plus 80 percent of the charges in the private sector. For the most part private insurance companies will pay both for head and body, and for hospital and private treatment. Perhaps the most authoritative evaluation of this problem was the report by the Institute of Medicine which was commissioned by Blue Cross. The interested reader is advised to familiarize himself with this document.[2]

Tables 9 and 10 are from an article by Evens and

Table 9. Typical revenues per patient*

	Gross ($)	*Net ($)†*
Technical		
Head without contrast	160	140.48
Head with and without contrast	214	187.89
Body without contrast	188	165.06
Body with and without contrast	222	194.92
Professional		
Head with contrast	54	47.41
Head with and without contrast	67	58.83
Body without contrast	64	56.19
Body with and without contrast	74	64.97

* From Evens and Jost: Radiology, March, 1978.

† Utilizing partial pay-bad debt reduction of 12.2%.

Table 10. Data comparison on head CT and body CT units*

	HCT	*BCT*
Hospital location (%)	89	89
Average hospital bed size	595	542
Radiologist responsible—head (%)	94	97
Radiologist responsible—body (%)	—	100
Installation time (months)	1.3	2.0
Maintenance time (hrs/week)	7.1	6.0
Usual operation (days/wk)	5.4	5.2
Usual operation (hr/day)	11.8	10.0
Patients examined/week	58	32
Head exams (%)	100	59
Studies with contrast—head (%)	60	68
Studies with contrast—abdomen (%)	—	65
Scheduling delay—inpatients (days)	1.6	0.9
Scheduling delay—outpatients (days)	11.5	3.4
Scheduling delay—increasing—inpatients (%)	21	14
Scheduling delay—increasing—outpatients (%)	35	18
Equipment cost ($)	387,000	526,000
Space allocated (ft^2)	537	694
Remodeling costs ($)	19,000	55,000
Annual cost of space ($/$ft^2$)	11.60	13.52
Technical personnel	4.6	3.0
Technical cost/study—50 patients/week ($)	130	151
Technical cost/study—30 patients/week ($)	201	234
Average partial pay-bad debt (%)	12.2	12.2
Average net technical revenue/patient	138	177

* From Evens and Jost: Radiology, March, 1978.

Jost[7] in part presented at The 1977 RSNA and kindly made available to me by the authors. Table 9 summarizes typical charges from their survey and assumes a bad debt reduction of 12.2 percent. Table 10 is a comprehensive comparison of head and body CT data.

Acknowledgments

The author wishes to greatfully acknowledge the invaluable assistance of Ms. Kathleen A. Ewing, the support of Radiology Consultants, P.C. during the preparation of this manuscript, and the opportunity provided by TSA to study this problem.

References

1. Better Health Care for Less Cost: The Case for Diagnostic Imaging and CT Scanning, Tehnicare Corporation Annual Report 1977. Ohio, Technicare Corp., 1977
2. Computed Tomographic Scanning: A Policy Statement, Institue of Medicine. Washington, D.C., National Academy of Sciences, Apr 1977
3. CT in Perspective: Diagnostic Efficacy and Health Care Implementation. Chicago, Illinois, EMI Medical Inc., Aug 1977
4. Effect of Computed Tomography on Hospital Practice. Northfield, Ill., J. Lloyd Johnson Associates, Dec 1976
5. Evens RG, Jost RG: Economic Analysis of Computed Tomography Units, Am J Roentgenol, 127:191, 1976
6. Evens RG, Jost RG: The Clinical Efficacy and Cost Analysis of Cranial C.T. and the Brain Scan, Sem Nucl Med, 7:129, 1977
7. Evens RG, Jost RG: Economic Analysis of Body Computed Tomography Units, Radiology, March 1978
8. Fineberg HV, Parker GS, Pearlman LA: CT Scanners: Distribution and Planning Status in the United States, N Engl J Med, July 28, 1977
9. Freedman, GS: C.T. in CT. or CATS in CONN. or Computerized Axial Tomography in Connecticut, Conn Med, 40(11), Nov 1976
10. Freedman, GS: Computerized Tomography-Development and Impact on Nuclear Medicine, Other Health Services, and Medical Economics. In Financial Operation and Management Concepts in Nuclear Medicine. Baltimore, University Park Press, 1976
11. Marshall, E: Anatomy of Health Care Costs: III, CAT Fever, New Repub, 176 (16), Apr 16, 1977
12. Shauffer, IA: Regulatory Activities of Local and State Health Services, Appl Radiol, Jan–Feb, 1977

Financial Planning and Economic Aspects of Acquiring Radiology Department Equipment*

Richard J. Oszustowicz

The purpose of this paper is to present a comprehensive financial analysis process by which a hospital and/or other health care organization manager, and particularly a radiology department head, could display the total financial requirements associated with the acquisition of major equipment for the radiology department. Note that this paper will therefore address total financial requirements, not just an element thereof: for example, should the equipment be purchased or leased; or how much working capital—front end monies—are required in order to introduce this new equipment into the health care organization.

In this paper a CT System acquisition is the equipment acquisition model against which total financial requirements are identified. Why a CT System as the equipment acquisition model for this paper? Well, the CT System requires more detailed discussion of total financial requirements than any other radiology department equipment acquisition being currently considered by radiology departments throughout the nation. Therefore, the reader of this paper will have received a financial planning model that is useful in the development of a three-year forecast for such capital equipment acquisitions in hospitals as legislated under current Medicare regulations. The eight-step financial planning process displayed in this paper is also useful for any radiology equipment acquisition irrespective of organizational affiliation, for example, for a private radiology practice and/or one associated with a private medical group practice.

In the opinion of the author, productivity forecasts, service cost forecasts, supply costs, depreciable lives, etc., are equipment and health care organization specific; consequently, the reference below to the manufacturer and to a 250-bed hospital (this paper means to introduce a realism that would be confused if reference was made only to the ABC Company and a medium size community hospital).

Exhibit 1 presents a step-by-step process by which the health care organization arrives at total financial requirements. The comprehensive three-year forecast also helps a Medicare participating hospital (and most hospitals are Medicare participating) in complying with Medicare Administrative Bulletin No. 234, dated May 1977, which states that a "separate cost center for the CT scanner should be established on the Medicare Cost Report. Establishment of this cost center would depend upon the provider's physical and accounting capability to meet current program criteria for identification of a cost center."

The CT System Financial Requirements exhibit aids a hospital toward the establishment of CT scanner services as a distinct cost center. Recall when nuclear medicine was embedded in radiology. Nuclear medicine merged as a distinct and viable department. A separate cost center was developed for nuclear medicine. CT scanner services are emerging into a cost center much like nuclear medicine did. The CT System Financial Requirements exhibit aids a hospital, or for that matter any health care organization, in establishing a distinct cost center.

Note that Exhibit 1 refers to total financial requirements, not cost/income analysis, cost analysis, or economic analysis. In 1969 the American Hospital Association published its current policy statement titled "Statement on Financial Requirements of Health Care Institutions and Services." A major purpose of that statement was to identify the total financial requirements that any health care organization should address in its consideration of any provided or to be provided health care service. The CT System Financial Requirements exhibit demonstrates a recognition and respect

*Reprinted from Applied Radiology, Vol. 7, No. 6, November/December 1978 Copyright by Barrington Publications, Inc. 825 S. Barrington Ave., Los Angeles, California 90049. Exhibit 2 is reproduced from: A Policy Statement: Computed Tomographic Scanning (1977) pp. 15–21 with the permission of the National Academy of Sciences, Washington D.C.

for this AHA financial policy statement, therefore assuring the health care organization that financial requirements are being viewed comprehensively by the manufacturer, in this case Searle CT Systems, a division of Searle Diagnostics, Inc., and the provider, the 250-bed hospital.

Exhibit 1 is supplemented by Exhibit 2, which presents certain excerpts from the National Academy of Sciences Institute of Medicine's April 1977 Policy Statement on Computer Tomographic Scanning. The excerpts selected concentrate on the efficacy of head scanning and body scanning. The Institute's statements clearly reflect a growing respect for the diagnostic accuracy and treatment management benefits of computed tomographic scanning. The Institute's representations may be beneficial for expert reference inclusion in support of patient demand forecasts in the financial impact section of Certificate of Need applications.

In order to present a most comprehensive paper, the author selected an actual 250-bed hospital radiology department as the health care organization that would display the total financial requirements associated with the introduction and operation of CT equipment into its radiology department. The equipment to be acquired by this 250-bed hospital is a Searle CT System, hereinafter referred to as the CT System. The author required the equipment to be identified by manufacturer at the opening of this paper in order to best validate production forecasts, service costs, supply costs, depreciable lives, etc.

Exhibit 1: CT System Financial Requirements

The following exhibit displays a health care organization's process of forecasting the total financial requirements associated with the acquisition of a CT System. The health care organization in this exhibit is a non-profit 250-bed hospital. Of course the size of the hospital is not necessarily the most important evidence supportive of CT acquisition. In this case, the hospital medical staff identified sufficient demand for CT diagnostic procedures.

In the instance of this paper, the inpatient and substantial outpatient demand for quality CT diagnostic procedures resulted in the selection of a Searle CT System. The diagnostic procedure demand for CT System services could just as well have been identified in larger or smaller health care organizations, for example—hospitals, ambulatory care facilities, medical group practices, etc. Such demand is evidenced in the form of substantial national and international purchase orders for new additions, replacement, upgrading, and/or expansion of existing CT services.

The purpose of this exhibit is to forecast the total financial requirements for the demand identified by the health care organization. The process flows through the following eight steps:

Step 1—Detail demand for CT diagnostic procedures, equipment, and site preparation costs
Step 2—Detail direct expenses
Step 3—Detail indirect expenses
Step 4—Detail working capital requirements
Step 5—Detail capital requirements
Step 6—Detail earning (profit) requirements
Step 7—Detail deductions from patient revenues and calculate gross patient revenues
Step 8—Detail provision for Federal, state, or local income taxes

Step-by-step development flows as follows:

Step 1: Detail Demand for CT Diagnostic Procedures, Equipment, and Site Preparation Costs

(a) Detail demand for diagnostic procedures to be performed by the CT System.
(b) Detail CT System equipment configuration.
(c) Detail site preparation costs.

Proceed to Steps 2 through 8. Identify the total financial requirements (resources) necessary to appropriately finance the demand for such diagnostic procedures.

Step 2: Detail Direct Expenses

Direct expenses are the costs of any goods or service that contribute to and are readily ascribable to the product or service output of the organization by which they are incurred.

Step 3: Detail Indirect Expenses

Indirect expenses are costs not readily identifiable with or incurred as a result of the production of specific goods or services, but applicable to a productive activity generally. The process by which indirect expenses are allocated in hospitals is identified as cost finding. Cost finding is the process of apportioning or allocating the cost of indirect expense, also classified as non-revenue-producing cost centers, on the basis of the statistical data that measure the amount of service rendered by each indirect expense cost center to other indirect expense and/or direct expense centers. In the instance of this exhibit the hospital cost finding process is designated by Medicare cost reporting requirements. In other words, the Medicare cost reporting process identifies the methodology by which indirect expenses are to be allocated for Medicare reimbursement purposes. Since the Medicare cost finding methodology represents how Medicare will pay for indirect expenses and since the Medicare patient represents a major patient workload for the CT System at this hospital, this indirect expense allocation process is judged to be a reasonable overhead allocation methodology for purposes of determining total financial requirements.

Step 4: Detail Working Capital Requirements .

Working capital is identified as capital in current use in the operation of the 250-bed hospital. It is the hospital's, and for that matter any business' investment in short-term or current assets, namely, cash, accounts receivable, and inventories.

For purposes of this exhibit, working capital is more specifically identified as the amount of dollars required to "front end"—underwrite—CT services rendered until cash inflows begin. Cash inflows occur from the flow of services rendered as identified by the evolution of the service and/or product through accounts receivable into cash. Note that CT services are usually identified as emerging from an established radiology and/or nuclear medicine service, therefore, in most situations the necessary front end (working capital) financing requirements are already in place, provided by prior overall health care organization, in this instance, hospital earnings.

Table 1: CT-System Financial Requirements

Financial Requirements			Year 1	Year 2	Year 3
Step 1:					
(a) Demand for CT-system diagnostic procedures	Inpatient		1,260	1,300	1,200
	Outpatient		840	1,300	1,800
For purposes of this paper, the health-care organization forecasts the following volume for its first three years of operation*:			2,100	2,600	3,000
	CT head scan — enhanced		1,800	2,000	2,200
	CT scan — abdomen		200	300	300
	CT body scan		100	300	500
			2,100	2,600	3,000
(b) Detail CT-equipment description					
CT-system base price (Pho Trax 4000)		$545,000			
With 80-megabyte disk drive		15,000			
And a multiimaging-format imager		7,000			
Total equipment cost		$567,000			
(c) Detail site-preparation costs					
Approximately 550 sq ft of floor space are required. This space is divided between an examination room and a control room. Such remodeling costs are averaging $20,000, with reported ranges of $15,000-$45,000. For purposes of this paper, $35,000 is identified as site-preparation cost.					
Site-preparation cost		$ 35,000			
Step 2: Detail direct expenses					
Salaries — two registered radiologic technologists			$22,200	$23,754	$25,416
It should be noted that the health-care organization may have a reserve of registered-radiologic-technologist time. However, for purposes of this paper, it is assumed that these two radiologic technologists will shift from the current health-care-organization-employed technologist group. These registered technologists are assumed to have three years of experience. Their prior duties in the department will be met through hiring new personnel. The salary shown is that of the third-year registered radiologic technologists. Assume first-year individual salaries of $11,100, each increased by 7% in years 2 and 3, reflecting movement of employees into higher steps within the registered-radiologic-technologist job description.					

(Step 2 continued on p. 24)

* See Exhibit 2 for the National Academy of Sciences Institute of Medicine listing of areas of the body for which CT scanning may be indicated.

Step 2 continued

Fringe Benefits	**Approximate Forecasts of Annual Increases**	**Year 1**	**Year 2**	**Year 3**
Federal Insurance Contributions Act (FICA)	7%	$1,287	$1,390	$1,486
Medical and hospitalization insurance	15%	718	844	993
Group life insurance	6%	72	78	84
Workmen's compensation	10%	504	550	605
Pension	15%	800	941	1,107
Total fringe benefits		$3,381	$3,803	$4,275

Note: Fringe-benefit increases are reflective of health-care-organization 1977 industry trends.

	Year 1	Year 2	Year 3
Physician Fees	$126,000	$156,000	$180,000

Fees charged by physicians for the supervision and interpretation of CT scanning examinations range from $25 to $75; the average professional fee is estimated at $55. For purposes of this paper, $60 per procedure is forecast. It should be noted that in some health-care organizations, the physician's professional fee would be privately billed to the patient by the physician. This paper assumes that the health-care organization would assume the billing responsibility to the patient on behalf of the physician. If the physician billed privately, $60 would not be included in the health-care organization's prices.

	Year 1	Year 2	Year 3
Supplies Expense	$20,622	$27,378	$33,840

	Supply Expense per Patient		
	Year 1	**Year 2**	**Year 3**
Contrast material	$4.00	$4.29	$4.60
Sheet film	3.50	3.75	4.02
Syringes	.86	.92	.98
Needles	.06	.07	.07
Venatube replacement	.40	.43	.46
Data-recording media	1.00	1.07	1.15
Total	$9.82	$10.53	$11.28

Note: A 7.20% supplies-expense increase is forecast for years 2 and 3.

	Year 1	Year 2	Year 3
Service Contract	—	$25,000	$26,500

First-year service needs are covered within the first-year warranty. Years 2 and 3 forecast $25,000 and $26,000 service arrangements, reflecting a 6% increase of Year 3 over Year 2.

	Year 1	Year 2	Year 3
Tube-Replacement Amortization	—	—	$2,666

Due to the manufacturer's CT technology, tube replacement usually forecast to occur in Year 2 is not forecast to occur until the end of the first quarter of Year 3. A cost of $8000 is forecast to be amortized over the remaining three years of the five-year depreciable life of the equipment.

	Year 1	Year 2	Year 3
Equipment Upgrading/Addition	—	—	—

No major equipment upgrading is forecast during the five-year depreciable life of the equipment. Demand shifts and/or technology changes may support equipment upgrading and/or addition sometime during the five-year depreciable life of the equipment. However, in its planning phase the health-care organization does not see any evidence that equipment upgrading or addition will occur in the CT-system equipment's five-year depreciable life.

	Year 1	Year 2	Year 3
Computer-Programming Upgrading	—	$6,250	$6,250

Diagnostic-procedure demand shifts and additional data needs are forecast to require some updating of computer programs. During the five-year period, $25,000 is forecast for such work. Such programming effort is forecast to begin in Year 2.

	Year 1	Year 2	Year 3
Equipment Depreciation	$113,400	$113,400	$113,400

The health-care organization has selected a five-year depreciable life for the equipment. Medicare Administration Bulletin No. 234, May 1977, supports a five-year depreciable life. Administrative Bulletin 234 read as follows: "The depreciation for the Computerized Tomography (CT) equipment should be based on a useful life of not less than five years." This useful life has been determined based on input (to the Bureau of Health Insurance) from industry consultants. No useful life has been published by the Internal Revenue Service; however, a survey conducted by the American Hospital Assn. indicated that the majority of hospitals use five-year straight-line depreciation for their CT equipment. Accelerated methods of depreciation may be justified. Medicare will allow 150% declining-balance depreciation upon proper presentation of the need to accelerate. Other third-party payers such as Medicaid and Blue Cross may also allow accelerated depreciation. For purposes of this paper, a five-year straight-line depreciation is determined appropriate by the health-care organization.

Also note that the hospital may choose to lease the equipment. Lease rates vary depending on the credit of the health-care organization. Lease arrangements should be analyzed against FASB Statement No. 13 to determine if the lease has to be capitalized, thereby requiring depreciation expense and interest expense to be recorded in lieu of lease expense. Also, if a lease is to be considered, the health-care organization should determine if a property tax will be assessed. In many states, not-for-profit as well as for-profit health-care organizations are subject to property taxes on leased equipment. Such expense would be treated as a direct expense.

	Year 1	Year 2	Year 3
Interest Expense	$51,030	$40,824	$30,618

The health-care organization was able to borrow the full $567,000 from its local bank for a five-year period at a 9% interest rate.

	Year 1	Year 2	Year 3
Site-Preparation-Costs Amortization	$1,750	$1,750	$1,750

Site-preparation costs for the 550-sq-ft area were forecast at $35,000. In accordance with generally accepted accounting principles, such building improvement should be amortized over the remaining useful life of the building that houses the CT-system service. A 40-year depreciable life is most prevalent for hospital buildings. For purposes of this paper, it is assumed that 20 years of depreciable life remain for the building in which the CT system is housed.

Under Medicare cost-reporting processes, site-preparation-amortization costs may not be charged directly to the CT-system cost center. It is customary to allocate building-remodeling costs (site-preparation costs) to all patient-revenue and non-patient-revenue departments on the basis of departmental square feet. Consequently, if the CT cost center occupies only 0.32% of the health-care organization's 170,000 sq ft, only about $6 would be allocated to the CT cost center for such site preparation under Medicare cost-reporting purposes. However, for purposes of this paper, and particularly for purposes of designing a CT-system diagnostic-procedure price to pay for all financial requirements associated with the equipment's acquisition, the site-preparation costs are identified as a direct cost of the CT cost center.

	Year 1	Year 2	Year 3
Insurance Expense	$1,350	$1,620	$1,944

Fire and general-liability coverage and malpractice coverage vary depending on the health-care organization's experience. Assume an annual price increase of 20%.

	Year 1	Year 2	Year 3
Total Direct Expenses	$339,733	$399,779	$426,659

Step 3: Indirect Expenses

The Medicare cost report provides an indirect-cost allocation that is usually identified as the step-down cost-finding process. Reference to the hospital's Medicare cost report disclosed the following indirect costs and their statistical bases of allocation.

Indirect Cost		Statistical Allocation Base	Unit Cost Multiplier	Year 1	Year 2	Year 3
Building Depreciation				$1,225	$1,225	$1,225
		Square footage: 550 sq ft	Building-Depreciation Expense per Square Foot			
		550 sq ft	$2.23			
Employee Health Services		Salaries	$0.0117	$260	$277	$297
Telephone		Number of extensions 2 of 350 extensions 0.5714%	Telephone expense: $195,629	$1,117	$1,175	$1,237
Telephone expense — asume a 5% annual increase						
Purchasing Department				$773	$1,062	$1,339
Purchasing-department costs — assume a 9% annual increase						
		Ratio of Dollar Value of CT-Supplies Expense to Total-Supplies Expense	Supplies			
Year 1 — $101,798	x	(20,622/2,687,000)0.76%				
Year 2 — 111,865	x	(27,378/2,880,464)0.95%				
Year 3 — 122,928	x	(33,840/3,087,857)1.09%				
Admitting Department				$1,188	$1,687	$2,117
Admitting-department costs — assume a 9% annual increase						
		Ratio of Inpatient Forecast CT Charges to Total Inpatient Charges				
Year 1 — $144,911		x	0.82%			
Year 2 — 159,242		x	1.06%			
Year 3 — 174,991		x	1.21%			
Business Office				$6,336	$9,947	$12,570
Business-office expense — assume a 9% annual increase						
		Ratio of Dollar Value of Forecast CT Total Patient Charges to Total Charges				
Year 1 — $452,600		x	1.40%			
Year 2 — 497,362		x	2.00%			
Year 3 — 546,552		x	2.30%			

	Statistical Allocation Base	Unit Cost Multiplier	Year 1	Year 2	Year 3
Administration and Personnel Office	Number of Full-Time-Employee Equivalents to Total Full-Time-Employee Equivalents — 2 employees ÷ 800 employees		$2,742	$2,988	$3,284
Year 1 — $1,096,831		0.25%			
Year 2 — 1,195,545		0.25%			
Year 3 — 1,313,786		0.25%			
Laundry	Pounds of laundry	Cost per Pound	$917	$1,188	$1,449
Year 1 — 924 pounds of CT laundry		$ .99			
Year 2 — 1,143 pounds of CT laundry		1.04			
Year 3 — 1,318 pounds of CT laundry		1.10			
Operation of Plant (heat, light, power, air conditioning)	Square feet	Operation of Plant Cost per Square Foot	$2,849	$3,129	$3,437
Year 1 — 550 sq ft		$5.18			
Year 2 — 550 sq ft		5.69			
Year 3 — 550 sq ft		6.25			
Housekeeping	Cleaning hours per year	Direct and Indirect Cost per Hour	$ 208	$ 229	$ 252
Year 1 — 36 hours per year		$5.80			
Year 2 — 36 hours per year		6.37			
Year 3 — 36 hours per year		7.00			
Cafeteria			$ 533	$ 586	$ 644
Cafeteria cost — assume a 9% annual increase	CT Service Full-Time Equivalent Employees to Total Full-Time Equivalent Employees — 2 employees ÷ 800 employees				
Year 1 — $213,373		0.25%			
Year 2 — $234,475		0.25%			
Year 3 — $257,665		0.25%			
Total Indirect Expenses			$ 18,148	$ 23,493	$ 27,851
Note that other indirect costs may be identified by a health-care organization. Such other indirect costs may reflect overhead costs peculiar to the organization. Usually such other indirect costs will be found in appropriate Medicare and other third-party-payer cost-reporting schedules.					
Total Direct and Indirect Expenses			$357,881	$423,272	$454,510

Note that working capital requirements are permanently needed. Consequently, working capital must be permanently included within the price of any product or service. The cash outflow cycle will always be waiting for future inflows simply because the nature of health care organization services creates patient accounts receivable for services rendered.

Step 5: Detail Capital Requirements

Capital is defined as those forecast upgrading and/or replacement costs attributable to land, building, and equipment associated directly or indirectly with future services. In the instance of the hospital identified in this exhibit, capital will be equal to the amount of dollars

	Year 1	Year 2	Year 3

Step 4: Detail Working-Capital Requirements

The health-care organization assumes that it will take 50 days to collect direct and indirect expenses from patient billings to price-paying and cost-paying third-party payers. In other words, the health-care organization must assume the financing of the CT-system service for 50 days until first cash inflow occurs. It should be noted that many health-care commercial insurers, Blue Cross/Blue Shield plans, and Medicaid programs are paying for CT scans. CT head scans are being paid for by the Medicare program. Medicare is currently reviewing its CT-body-scan reimbursement policy. Currently, Medicare patients are privately paying for CT body scans that are not recognized for payment by the Medicare program.

Assume that the working-capital need is equal to 50 days/365 days or 14% of direct and indirect expenses. In other words, the health-care organization must assume the financing of the CT-system service for 50 days until first cash inflow occurs. The 50 days is assumed to be net of any working-capital financing through accounts payable owing suppliers and other short-term creditors. Note that this 50-day working-capital requirement is a permanent working-capital need, since all billings are assumed to take 50 days to collect for all future billing periods. Consequently, even though the radiology department and/or the health-care organization may have the necessary equity to finance the initial 50-day collection/waiting period, it must nevertheless price for such in order to assure a permanent funding of this very important financial requirement — working capital.

For purposes of this presentation, assume that the radiology department borrowed its first year's working-capital requirement of $50,103 from restricted hospital funds. Consequently, two years of working-capital pricing are required to secure the necessary equity for permanent establishment of this important financial requirement. In both years 2 and 3, additional working capital of $9,155 ($59,258 less $50,103) and $13,528 ($63,631 less $50,103) was required to cover incremental working-capital needs due to volume changes and increased costs. Assume that in Year 3 the radiology department forecasts a significant expansion of working-capital requirements due to increased volume forecasts in Year 4 as well as to increased state rate-setting regulation, which will again force debt financing of Year 3 working-capital needs.

		Total Direct and Indirect Expenses	
Year 1 — 14%	x	$357,881	=
Year 2 — 14%	x	423,272	=
Year 3 — 14%	x	454,510	=

	Year 1	Year 2	Year 3
Working Capital	$50,103	$59,258	$63,631

	Year 1	Year 2	Year 3

Step 5: Detail Capital Requirements

The health-care organization indicated that it would pursue a pricing plan to enable it to place 25% of the current purchase price of the CT system ($567,000) or $28,350 annually over the five-year period toward the financing of upgrading or replacement costs that may be associated with the equipment in the future. As represented above, the organization had obtained 100% financing for the current acquisition at 9% interest. The organization has established a goal of funding $28,350 annually to enable it to best compete in future financing markets.

	Year 1	Year 2	Year 3
Capital	$28,350	$28,350	$28,350

Step 6: Detail Earnings (Profit) Requirements

The health-care organization identifies earnings (profit) as the appropriate source of financing future working-capital expansion associated with the CT system. Note that direct and indirect expenses increase $65,391 and $31,238 in years 2 and 3 over the previous year. Such working-capital expansion will have to be developed in prior-period prices in order to be available at the beginning of a succeeding fiscal period. Additional earnings (profit) are required to finance bad-debt expense, third-party-payer contractual adjustments, and other deductions from patient revenue. Earnings (profit) provide the necessary cash for future-service development. This health-care organization identifies an earnings (profit) requirement in absolute dollars of $72,622 over the first three years of the CT service.

	Year 1	Year 2	Year 3
Earnings (profit)	$ 452	$20,786	$ 51,384

Step 7: Detail Deductions from Patient Revenues

	Year 1	Year 2	Year 3
Deductions from Patient Revenues	$67,214	$92,334	$122,125

Deductions from patient revenues identify the dollars of price (revenue) that will not be paid because of third-party-payer contractual arrangements. For example, Medicare, Medicaid, and many Blue Cross/Blue Shield plans pay direct and indirect expenses only and will therefore not contribute to capital and earnings requirements. Additionally, the hospital or other health-care organization will experience bad-debt write-offs and charity allowances. In this paper the health-care organization assumes the following patient-CT-revenue volume, which will impact on deductions from patient revenues.

				Year 1	Year 2	Year 3
Excess of gross patient revenues over operating expenses				$146,119	$200,728	$265,490
Blue Cross/Blue Shield	12.8%	x	Excess of	18,705	25,696	33,988
Medicare	21.6%	x	gross patient	31,566	43,362	57,351
Medicaid	3.3%	x	revenue over	4,819	6,620	8,756
Bad-debts expense	8.0%	x	operating	11,688	16,056	21,237
Charity services	0.3%	x	expenses	436	600	793
Total	46.0%	x		$ 67,214	$ 92,334	$122,125

In other words, 46.0% of the health-care organization's excess of gross patient revenue over direct and indirect expense will not be experienced. Conversely, 54.0% of all revenue is assumed to be billed to paying health-care consumers, who therefore finance the capital and earnings requirements of the organization. The deductions-from-revenue experience of this organization reflect the usual situation in the nation; such deductions vary by degree as they are influenced by the size and mix of the components of deductions from patient revenue.

(Step 7 continued on p. 30)

the hospital perceives should be earned so as to enable it to prudently finance future upgrading and/or replacement of the CT System. Additional equipment incremental to initial CT System installation is forecast to be financed through earnings (profit), as further described in Step 6.

Step 6: Detail Earning (Profit) Requirements

Profit is defined as the excess of price over all direct and indirect expenses, bad debts expense, third-party payer contractual adjustments, and other administrative adjustments to patient revenue. It is the reward to a business institution, profit and nonprofit health care organization, and/or person(s) for the risks assumed in the establishment, operation, and management of a given enterprise or undertaking. It is a major source of cash for working capital expansion associated with future higher demand for CT services—equipment expansion, departmental or overall health care organization research and development of new techniques and procedures to assure effective, efficient, and economical delivery of health care services.

Step 7: Detail Deductions From Patient Revenues and Calculate Gross Patient Revenues

Deductions from patient revenues include:

(a) Allowances which represent differences between gross patient revenue charges and amounts received (or to be received) from patients or third-party payers. Types of allowances are:

(1) Charity allowances—the difference between gross revenue charges at established rates and amounts received (or to be received) from indigent patients, voluntary agencies, or governmental units on behalf of specific indigent patients.

Step 7 continued

Calculation of Gross Patient Revenue

	Year 1	Year 2	Year 3
Inpatient	$302,400	$312,000	$288,000
Outpatient	201,600	312,000	432,000
Total	$504,000	$624,000	$720,000

The National Academy of Sciences Institute of Medicine in its April 1977 policy statement on computed tomographic scanning stated that "charges for the technical component of CT examinations vary from $100 to $440, with an average of $240;* the charge for a study with contrast media — used for image enhancement in 40%-60% of scans† — averages $67. About 10% of all institutions charge a standard fee for CT scanning, regardless of whether the scan is unenhanced, enhanced, or both. The others make separate charges for enhanced and unenhanced scans and may have a third rate when both unenhanced and enhanced scans are ordered simultaneously."

Upon analysis of its total financial requirements, the health-care organization in this paper developed an average CT-diagnostic-procedure price of $240, which includes the physician's professional component. The price is assumed not to increase over the forecast three-year period due to the economies-of-scale impact in future years on major fixed costs. Also, it should be noted that a federal and/or state cost-containment program could be manageable within the $240 established three-year price.

Calculation of Gross Patient Revenues

	CT Diagnostic Procedures				
Year 1 —	2,100	x	$240	=	$504,000
Year 2 —	2,600	x	$240	=	$624,000
Year 3 —	3,000	x	$240	=	$720,000

Note that outpatient revenue is forecast at 40%, 50% and 60% for years 1, 2, and 3, respectively. The 60% outpatient volume is expected to be sustained through the 5th year.

* American Hospital Association: CT Scanners: A Technical Study, p. 92

† Evens RG, Jost RG: Economic analysis of computed tomography units, Am J Roentgenol 127: 197, 1976

(2) Courtesy allowances or policy discounts—the difference between established rates and amounts received for services provided for doctors, clergymen, employees, and employees' dependents and others as may be determined.
(3) Contractual allowances—the difference between established rates and amounts received or to be received from third-party payers under contractual arrangements. This allowance recognizes that third-party payers, such as Medicare, Medicaid, and Blue Cross/Blue Shield organizations, do not pay hospitals and other health care organizations published prices, but usually pay only direct and indirect costs associated with the service rendered.

(b) Provision for uncollectible accounts—an allowance for bad debts expense. This allowance recognizes that certain patients will simply not pay their medical bills. These patients would have received medical services, the health care organization would have expended financial resources in the provision of care to that patient, with the patient simply not paying all or a portion of the price charged.

Note that these deductions from patient revenues are deductions from price charged. The total (five) financial requirements identified in Steps 2 through 6—namely, direct costs, indirect costs, working capital, capital, and earnings—must still be met in order to maintain and preserve the financial integrity of the health care organization. In other words, the prices charged all patients are established in such a way that although some patients will be granted certain price deductions as described above, other patients, by paying the published price, will assure the health care organization that its total financial requirements will be met. It is clear that such allowances exist in all forms of profit and nonprofit organizations.

Step 8: Detail Provision for Federal, State, or Local Income Taxes

Profit health care organizations will have to establish prices which recognize federal, state, and local income taxes. Such taxes are assessed against the financial requirements identified as earnings (profit). Profit goals will vary by organization. Such earnings provide a source of funds for federal, state, or local income taxes. The profit organizations' goals for net earnings (net profit after provision for such income taxes) will of course influence the price established for CT System diagnostic procedures.

Step 8: Provision for Federal, State, and Local Income Taxes

This health-care organization is a nonprofit organization. Consequently, Step 8 is not applicable. However, it should be noted that the eight-step flow of this system arrives at an earnings (profit) amount against which the provision for federal, state, and local taxes could be calculated.

A Summary of CT-System Financial Requirements

Financial Requirements		Year 1	Percent	Year 2	Percent	Year 3	Percent
Step 1:	Demand for CT-system diagnostic procedures	2,100		2,600		3,000	
	Detail Searle CT equipment description	$567,000		—		—	
	Detail site-preparation costs	$ 35,000		—		—	
Forecast CT-system diagnostic procedures demand financial requirements							
Step 2:	Detail direct expenses						
	Salaries	$ 22,200	6.21	$ 23,754	5.62	$ 25,416	5.60
	Fringe benefits	3,381	0.94	3,803	0.89	4,275	0.94
	Physician fees	126,000	35.21	156,000	36.85	180,000	39.62
	Supplies expense	20,622	5.77	27,378	6.47	33,840	7.45
	Service contract	—	—	25,000	5.91	26,500	5.84
	Tube-replacement amortization	—	—	—	—	2,666	0.58
	Equipment upgrading/addition	—	—	—	—	—	—
	Computer-programming upgrading	—	—	6,250	1.47	6,250	1.37
	Equipment depreciation	113,400	31.69	113,400	26.80	113,400	24.94
	Interest expense	51,030	14.25	40,824	9.64	30,618	6.73
	Site-preparation-costs amortization	1,750	0.48	1,750	0.41	1,750	0.38
	Insurance expense	1,350	0.37	1,620	0.38	1,944	0.42
Total direct expense		$339,733	94.92	$399,779	94.44	$426,659	93.87

(Summary continued on p. 32)

Summary continued

		Year 1	Percent	Year 2	Percent	Year 3	Percent
Step 3:	Detail indirect expenses						
	Building depreciation*	1,225	0.35	1,225	0.29	1,225	0.28
	Employee health services	260	0.07	277	0.06	297	0.06
	Telephone	1,117	0.31	1,175	0.28	1,237	0.29
	Purchasing	773	0.21	1,062	0.25	1,339	0.30
	Admitting department	1,188	0.34	1,687	0.40	2,117	0.46
	Business office*	6,336	1.79	9,947	2.37	12,570	2.72
	Administration and personnel office*	2,742	0.77	2,988	0.70	3,284	0.74
	Laundry	917	0.25	1,188	0.28	1,449	0.32
	Operation of plant	2,849	0.80	3,129	0.75	3,437	0.77
	Housekeeping	208	0.05	229	0.05	252	0.05
	Cafeteria	533	0.14	586	0.13	644	0.14
Total indirect expense		$18,148	5.08	$23,493	5.56	$27,851	6.13
Total direct and indirect expenses		$357,881	71.00	$423,272	67.84	$454,510	63.11
Direct expenses brought forward		$339,733	67.40	$399,779	64.07	$426,659	59.25
Indirect expenses brought forward		18,148	3.60	23,493	3.77	27,851	3.86
Total direct and indirect expenses brought forward		357,881	71.00	423,272	67.84	454,510	63.11
Step 4:	Detail working-capital requirements	50,103	9.94	59,258	9.49	63,631	8.83
Step 5:	Detail capital requirements	28,350	5.62	28,350	4.54	28,350	3.94
Step 6:	Detail earnings (profit) requirement	452	0.09	20,786	3.34	51,384	7.16
Step 7:	Detail deductions from patient revenues	67,214	13.35	92,334	14.79	122,125	16.96
Total gross patient revenue		$504,000	100.00	$624,000	100.00	$720,000	100.00

*Note that these indirect expenses do not necessarily represent incremental cash requirements due to the installation of the CT-system equipment. The health-care organization's indirect-expense allocation process simply shifts these indirect expenses to a patient-revenue-producing center. In other words, these indirect expenses would have been incurred by the health-care organization irrespective of its decision to acquire CT equipment. Since the decision was made to acquire the CT-system equipment, these indirect expenses are spread to the CT department, thus affording an opportunity for other patient-care revenue-producing departments to contain and/or lower prices; these indirect expenses no longer have to be financed within their pricing structure. Conversely, the health-care organization may view these overhead shifts as an opportunity to finance necessary expenses in the business, administration, and personnel offices.

Exhibit 2: Excerpts from the Institute of Medicine Policy Statement on Computed Tomographic Scanning, April 1977

The following statements are excerpted from the April 1977 Institute of Medicine's Policy Statement on Computed Tomographic Scanning. The Institute of Medicine was chartered in 1970 by the National Academy of Sciences to enlist distinguished members of appropriate professions in the examination of policy matters pertaining to the health of the public. In this, the Institute acts under both the Academy's 1863 Congressional charter responsibility to be an adviser to the federal government, and its own initiative in identifying issues of medical care, research, and education.

The following statements represent the Institute's perception of the efficacy of CT head and body scanning. The statements are presented to support the view that CT scanning, particularly CT body scans, is moving closer to recognition as "efficacious at the level of diagnostic impact when used for specific indications limited largely but not entirely to diagnosis and managing treatment of cancers at certain sites." Such acknowledgment increasingly brings body scanning into payment recognition by major third-party payers such as Medicare, Medicaid, and Blue Cross. The areas of the body for which CT scanning is or is not indicated at the time of the Institute's April 1977 statement should help a health care organization identify where the specific demands will flow for such services from their medical staff.

The National Academy of Sciences Institute of Medicine presented the following statements regarding the efficacy of CT head and body scanning:

Efficacy of CT Head Scanning

Using current standards for evaluating clinical evidence, the committee finds CT scanning of the head to be efficacious at the level of diagnostic impact when used to diagnose and determine the effect of treatment

Summation	Year 1	Percent	Year 2	Percent	Year 3	Percent
The following three-year forecast statement of revenues and expenses (income statement) collects the financial requirements discussed on the previous pages.						
Gross patient revenue						
Inpatient	$302,400	60.00	$312,000	50.00	$288,000	40.00
Outpatient	201,600	40.00	312,000	50.00	432,000	60.00
	504,000	100,00	624,000	100.00	720,000	100.00
Less operating expenses						
Direct expenses	339,733	67.40	399,779	64.07	426,659	59.25
Indirect expenses	18,148	3.60	23,493	3.77	27,851	3.86
	357,881	71.00	423,272	67.84	454,510	63.11
Excess of gross patient revenue over operating expenses	146,119	29.00	200,728	32.16	265,490	36.89
Less deductions from patient revenues	(67,214)	(13.35)	(92,334)	(14.79)	(122,125)	(16.96)
Excess of gross patient revenue over operating expense less deductions from patient revenues†	78,905	15.65	108,394	17.37	143,365	19.93
Less contribution to						
Working capital	50,103	9.94	59,258	9.49	63,631	8.83
Capital	28,350	5.62	28,350	4.54	28,350	3.94
	78,453	15.56	87,608	14.03	91,981	12.77
Earnings (profit)	$ 452	0.09	$ 20,786	3.34	$ 51,384	7.16

†Note that traditional financial forecasts would identify the excess of gross patient revenue over operating expenses less deductions from patient revenues as an organization's earnings (profit). This schedule breaks the excess of gross patient revenue less deductions from patient revenue into the specific financial requirements being funded by that excess — namely, working capital and capital. Restated, the bottom line should not be viewed as purely earnings (profit), since so much of it is devoted to the maintenance of working capital and preservation of capital. Through this summation, the health-care organization is able to detail support for its CT-system diagnostic-procedure price by specifically identifying total financial requirements.

on mass and structural lesions in or about the brain, including the meninges and orbit. CT scanning holds promise in demonstrating other lesions, including those resulting from demyelinization and cerebritis. Indications for a scan should include sufficient clinical information for determining the area of the head to be scanned and whether contrast should be used. CT scanning of the head is not an appropriate diagnostic procedure in the absence of strong clinical indications and supporting signs and symptoms. CT scanning of the head is usually efficacious in comparison with other diagnostic procedures.[1] Cerebral angiography, radionuclide scans, pneumoencephalography, echoencephalography, and skull x-rays will continue to have diagnostic utility, but CT is likely to replace them to some extent, to judge from comparisons of the information and risk associated with each method. The committee supports the decision of most third-party payers to reimburse for CT head scanning when competent judgment finds it clinically indicated.

Efficacy of Body Scanning

Although many possible uses of CT scanning in the body are presently under research, recent studies have presented substantial evidence meeting current standards that CT scanning is diagnostically accurate in certain applications.[2] The committee has considered the most recent results obtained by clinical researchers and has applied its best judgment in evaluating these results. On this basis, the committee finds that CT scanning of the body is efficacious at the level of diagnostic impact when used for specific indications, limited largely but not entirely to diagnosis and managing treatment of cancers at certain sites. The areas of the body for which CT scanning is or is not indicated at this time are listed below.

Neck CT scanning is not indicated at this time.

Chest

PLEURA

Detection of pleural metastases and other chest wall lesions.

LUNG

Detection of multiple tumor nodules where one or more have been found by conventional x-ray techniques. If there is clearcut evidence of bilateral involvement, CT is not appropriate.

Search for a primary tumor when a positive sputum for malignant cells has been obtained, but no evidence has been found through conventional x-ray techniques.

Determination of extent of spread of tumor to adjacent lobes in patients with impaired pulmonary functions.

Differentiation of solid, cystic, fatty, inflammatory, and vascular masses.

CT is not indicated for detection of pulmonary emboli at this time.

MEDIASTINUM

Detection and evaluation of masses.

Determination of extent of primary or secondary tumor.

HEART

Studies of the heart are not indicated at this time.

Great Vessels (including abdominal aorta)

CT scanning is not indicated in the aorta and great vessels except in the few postoperative patients in whom aortic graft abscesses are suspected.

Spine and Contents

SPINAL CORD

CT is not indicated in the spinal cord at this time.

SPINAL COLUMN

Determination of content and extent of meningoceles and meningomyeloceles.

CT biopsies.

Otherwise, CT scanning of the spinal column is indicated only where other procedures, including conventional tomography, radionuclide scanning, and myelography have failed to detect primary tumors, metastases, and inflammatory diseases in the presence of persistent symptoms.

Abdomen

RETROPERITONEAL AREA

Diagnosis and staging of nodal and extranodal extension of lymphomas, determination of extent of retroperitoneal involvement with lymphomas, and extent of other types of retroperitoneal metastases from various primary sites.

Detection of primary malignancies such as those of mesenchymal, neural, lymphatic, embryonic rest origin, melanomas, and benign conditions such as cysts which may mimic malignancies.

PERITONEUM

Detection and aspiration of abscesses and cysts.

LIVER

Search for primary and secondary tumors and some life-threatening benign lesions such as liver cell adenomas and cavernous hemangiomas and abscesses.

Determination of extent of tumor and differentiation of solid, cystic, inflammatory, vascular, and fatty lesions.

CT biopsies.

SPLEEN

CT is not indicated at this time.

PANCREAS

Search for primary and secondary tumor.

Determination of extent of tumor.

Differentiation of solid, cystic, inflammatory, vascular, and fatty lesions.

CT biopsies.

KIDNEY

CT scanning of the kidney is indicated only when preceded by a conventional IVP study, and then for:

Search for primary and secondary tumor.

Determination of extent of tumor.

Differentiation of solid, cystic, inflammatory, vascular, or fatty lesions.

CT biopsies or aspiration.

GALL BLADDER

CT is not indicated at this time.

BILIARY TREE

Differentiation of obstructive from nonobstructive jaundice.

GASTROINTESTINAL TRACT

CT is not indicated at present except for determination of extent of tumor spread to other organs.

ADRENAL GLANDS

Search for primary and secondary tumor.

Determination of extent of tumor.

Differentiation of solid, cystic, inflammatory, vascular, or fatty lesions.

CT biopsies.

Pelvis

UTERUS AND OVARIES

CT scanning is indicated after detection of a mass by clinical examination or after positive biopsy for:

Evaluation of primary tumor and its extent of spread; and evaluation of secondary tumor.

Differentiation of solid, cystic, inflammatory, vascular, or fatty masses.

CT biopsies.

BLADDER, URETERS, AND PROSTATE

Evaluation of primary and secondary tumor, including extent of tumor.

Differentiation of solid, cystic, inflammatory, vascular, or fatty tumors.

CT biopsies.

FLAT BONES

Evaluation of bone lesions.

CT biopsies.

EXTREMITIES

CT is indicated for determining the local extent of a tumor and presence of regional metastases.

THERAPY PLANNING AND FOLLOW-UP

CT is indicated for collection of information on cross-sectional anatomy and attenuation coefficients of bone and soft tissue in tumor-bearing areas for planning surgery and radiation therapy.

Follow-up evaluation of effectiveness of radiotherapy, surgery, or chemotherapy in cancer patients at primary or metastatic tumor sites when part of an established and acceptable follow-up protocol or when signs and symptoms suggest progression, recurrence, or failure of therapy.

Foreign Body Localization in the chest and abdomen

Conditions for which CT scanning is more hazardous than or diagnostically inferior to other procedures were not included in the list of indications. For some indications listed, other tests may be more appropriate in particular patients. If other diagnostic tests have permitted a definitive diagnosis to be made, CT scanning is justified only for planning treatment. Conversely, if a CT scan establishes a definitive diagnosis, additional diagnostic tests are unjustified. Sometimes, tests may complement each other either by providing different information or when one test succeeds after the first has failed to yield useful information. Recent studies comparing CT scanning with ultrasonic imaging of the abdomen suggests the two methods are complementary.[3]

Based on current evidence, CT is not superior in all applications. For dynamic studies of the circulatory and digestive systems and for high-resolution radiography in which structural details below a millimeter must be discerned, CT cannot compete with conventional radiographic techniques. In mammography, for example, xero-radiography provides definitive diagnostic information at a lower cost, although at a higher radiation level. Ultrasonic imaging is safer and, therefore, diagnostically superior to CT scanning in obstetrics and gynecology. In cardiology, TM mode and real-time ultrasonic imaging provide more valuable data than do currently available CT scanners. CT scanning cannot replace those nuclear medical techniques that provide unique information about body functions and body chemistry, as in the case of thyroid scans.

Because CT scanning of the body is an efficacious diagnostic tool for the conditions listed above on the basis of current standards of evidence, the committee recommends that CT scanning of the body when used for appropriate indications be recognized as a covered service under third-party reimbursement plans until and unless a decision is made to require more demanding standards of evidence for these decisions. However, experience with body scanning is evolving rapidly and the list of indicators for which coverage is warranted should be reviewed at least every six months. Therefore, the committee recommends that:

CT scanning of both the head and body, when appropriately used for specified indications, should be a covered diagnostic service under third-party reimbursement plans, accepting as criteria of efficacy the usual standards of clinical practice.

A Concluding Observation

By addressing itself to the details of total financial requirements as influenced by deductions from patient revenues and where applicable, provision for income taxes, this paper represents to the reader its commitment to quality financial analysis. Such comprehensive financial analysis will provide the reader with a tested financial analysis process that will surely aid the health care organization in making a timely CT System equipment acquisition decision.

This financial analysis process will also aid the health care organization in its need to logically and comprehensively present the total financial requirements to internal (board of directors) and external (state and federal Certificate of Need agencies) organizations. The process is also useful in supporting CT diagnostic procedure prices before state health care organization rate review and rate setting agencies.

The reader should not confuse the technique of data collection (in this case an eight-step, three-year financial planning/forecasting process) with the use of the data so developed. The use of the data developed is what is important since its usage enables the radiology department head or any health care organization manager to make a timelier decision, thereby affording the health care consumer a timelier opportunity to receive appropriate medical care.

Acknowledgments

The author wishes to acknowledge Searle CT Systems, particularly Mr. Philip Shevick, Vice President of New Business Develpment; Mr. John Derr, Vice President of Imaging Marketing Products Group; and Mr. Richard Schmidt, Group Vice President, Searle Diagnostics, Inc., for their help and advice in the development of this paper. Searle CT Systems provided the author with actual experiences in order to comprehensively develop this paper.

References

1. Weinstein, MA, Alfidi RJ, Duchesneau P: Computed Tomography, Six—Skull Roentgenography, One, Am J Roentgenol (in press)
2. Alfidi RJ, Antunez AR, Haaga JR: Computed Body Tomography and Cancer, Am J Roentgenol 127:1061, 1976
3. Sample WF: Comparison of Diagnostic Ultrasound and CT Scanning in the Torso, presented seminar of Bureau Radiological Health, FDA, PHS, USDHEW, Rockville, Md., 12 Jan, 1977

See also:

Alfidi RJ et al.: Computed Tomography of the Liver, Am J Roentgenol 127:69, 1976

Axelbaum SP et al.: Computed Tomographic Evaluation of Aortic Aneurysms, Am J Roentgenol 127:75, 1976

Carter BL et al.: Unusual Pelvic Masses: A Comparison of Computed Tomographic Screening and Ultrasonography, Radiology 121:383, 1976

Chernak, ES et al.: The Use of Computed Tomography for Radiation Therapy Treatment Planning, Radiology 117:613, 1975

Gramiak R, Waag RC: Cardiac Reconstruction Imaging in Relation to Other Ultrasound Systems and Computed Tomography, Am J Roentgenol 127:91, 1976

Haaga JR, Alfidi RJ: Precise Biopsy Localization by Computed Tomography, Radiology 18:603, 1976

Haaga JR et al.: Definitive Treatment of a Large Pyogenetic Liver Abscess with CT Guidance, Cleveland Clin Quart 43:85, 1976

Haaga JR et al.: Computed Tomography of the Pancreas, Radiology 120:589, 1976

Haaga JR et al.: Definitive Role of CT Scanning of the Pancreas: Two Years' Experience, presented 62nd meeting RSNA, Chicago, 13–19 Nov, 1976

Hammerschlag SB et al.: Computed Tomography of the Spinal Canal, Radiology 121:361, 1976

Hattery RR et al.: Urinary Tract Tomography, Radiol Clin 14:23, 1976

Hessel SJ et al.: Computed Axial Tomography of the Ischemic Heart, presented 62nd meeting RSNA, Chicago, 13–19 Nov, 1976

Kreel L: The EMI Whole Body Scanner in the Demonstration of Lymph Node Enlargement, Clin Radiol 27:421, 1976

MacIntyre WJ et al.: Application of Computer Processing of Total Body CT Scans to Delineate Structures in Renal Masses, presented 62nd meeting RSNA, Chicago, 13–19 Nov, 1976

Philips RL et al.: Computed Tomography of Liver Specimens, Radiology 115:43, 1975

Raskin, MM, Cunningham JB: Complementary Role of Computed Tomography and Ultrasound in the Abdomen and Pelvis, presented 62nd meeting RSNA, Chicago, 13–19 Nov, 1976

Ruegsegger P et al.: Quantification of Bone Mineralization Using Computed Tomography, Radiology 121:93, 1976

Sagle SS et al.: Computed Tomography (EMI Scanner) of the Thorax, presented 62nd meeting RSNA, Chicago, 13–19 Nov, 1976

Sagle SS et al.: Computed Tomography in the Evaluation of Diseases of the Pancreas, presented 62nd meeting RSNA, Chicago 13–19 Nov, 1976

Sagle SS et al.: Computed Tomography of the Kidney, presented 62nd meeting RSNA, Chicago, 13–19 Nov, 1976

Sheedy PF: CT in Evaluation of Patients with Carcinoma of the Pancreas, presented 62nd meeting RSNA, Chicago, 13–19 Nov, 1976

Sheedy PF et al.: Computed Tomography of the Body: Initial Clinical Trial with the EMI Prototype, Am J Roentgenol 127:23, 1976

Stanley RJ et al.: Computed Tomography of the Body:Early Trends in Application and Accuracy of the Method, Am J Roentgenol 127:53, 1976

Stephen DH et al.: Computed Tomography of the Abdomen. Early Experience with the EMI Scanner, Radiology 119:331, 1976

Ter-Pogossian M et al.: Computed Tomography of the Heart, Am J Roentgenol 127:79, 1976

Wiggans G et al.: Computerized Axial Tomography for Diagnosis of Pancreatic Cancer, Lancet 2:233, 1976

COMPARATIVE IMAGING TECHNIQUES

Advances in Ultrasound

Gordon S. Perlmutter

The recent advances in ultrasound have basically been technological, with newer instrumentation resulting in increasing diagnostic accuracy and improved ability to image difficult areas of the body. These improvements have occurred in four general categories, including (1) gray-scale high resolution scan converters, (2) automated scanning arms, (3) real-time imaging, and (4) endoscanning.[1,2]

Gray-Scale Scan Converters

Prior to the advent of gray-scale scan converters, the only way possible to integrate the ultrasound signal while scanning a patient was to use an open shutter photographic technique or a bistable storage scope. Each of these two modalities had several limitations. While it was possible to obtain good quality gray-scale images by placing a camera with an open shutter in front of a cathode ray tube and leaving the shutter open through the entire scanning process, good quality scans were exceedingly difficult to obtain with any degree of consistency. Any variation in the scanning process, such as varying the speed of scan or overwriting a particular area of the anatomy, resulted in objectionable photographs. Furthermore, the operator was unable to observe the developing image and could only view the end product after the shutter was closed and the film developed.

The alternative to the photographic open-shutter approach used by all manufacturers of ultrasound scanners prior to the advent of gray-scale scan converters was the bistable oscilloscope. As the name implies, the bistable scope is able to store the sound signal in only two shades of gray, i.e., black and white. This results from the characteristics of the storage elements on the surface of the bistable scope, which when activated are fully lighted and when not activated are completely off. An advantage of this type of image storage is the fact that once an image element is lighted, it cannot be further affected by overwriting or inconsistencies in the motion of the scanning arm. Furthermore, the operator is able to observe the image being integrated as the scan progresses and is able to interact with the developing image. The disadvantage, of course, is the limited dynamic range of the storage scope. Also, the matrix of storage elements on the surface of the scope is rather coarse, resulting in scans of rather poor resolution.

With the advent of gray-scale scan converters around 1974, the use of bistable storage scopes was abandoned by all manufacturers and replaced with the newer gray-scale units. Gray-scale scan converters had all the advantages of the bistable storage scopes with none of the disadvantages. The first and most important advantage of the gray-scale scan converter is the much larger dynamic range of operation which results in the ability to reproduce echoes of varying strength in eight to ten shades of gray. There is also a marked improvement in the resolution of the stored image. An example of a bistable and gray-scale scan performed on the same patient in the upper abdomen is shown in Figure 1. Other advantages of the gray-scale scan converters include very rapid writing time and the ability to operate in the peak detection mode, which in essence results in any one point on the scan converter tube storing only the strongest signal presented to it during the scanning process, thus correcting for overwriting and inconsistent scanning motion. Another advantage of a scan converter is that it not only stores a cathode ray signal over a period of time, but also converts it into a standard 525-line commercial television format, allowing for the use of television monitors as well as a host of other peripheral devices developed for commercial television, such as TV character generators, videotape recorders, and television image processors. By operating simultaneously in the " read" and "write" modes, these scan converters allow the ultrasound technologist to observe and interact with the scan as it is being produced.

Until recently, all scan converters produced commercially were of the analog variety. A sketch of an analog scan converter tube is given in Figure 2. In operation, a stream of electrons corresponding in strength to the returning ultrasound signal is aimed at a silicon dioxide image screen, with the direction of the electron stream corresponding to the direction of the scanning arm at

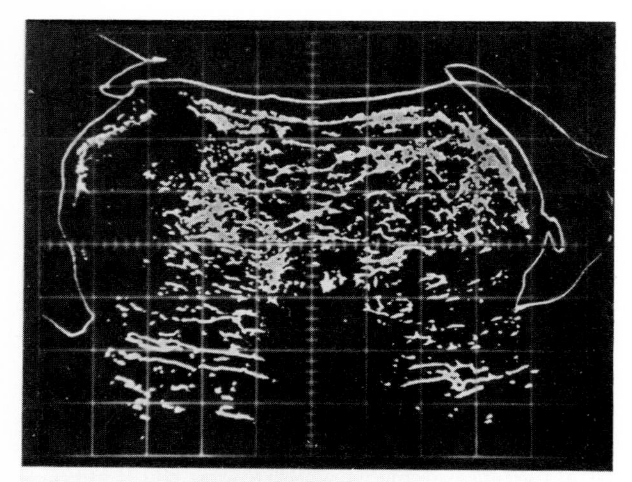

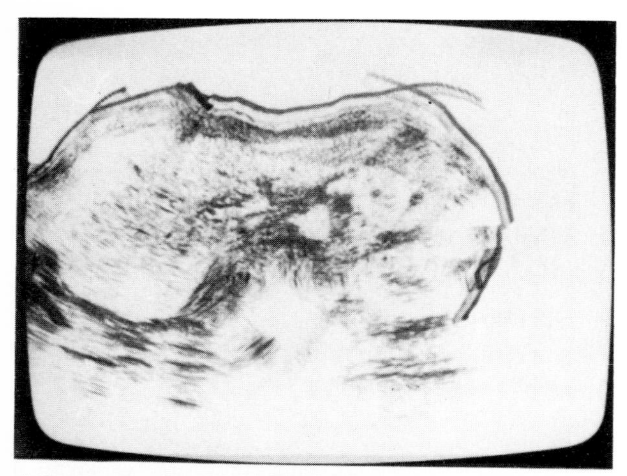

Figure 1. Examples of a transverse B mode scan of the upper abdomen are shown both as a bistable and as a gray-scale display, emphasizing the improved tonality and resolution of the gray-scale display. Reproduced from *Abdominal Gray Scale Ultrasonography,* B. B. Goldberg (ed), by permission of John Wiley & Sons, New York, 1977.

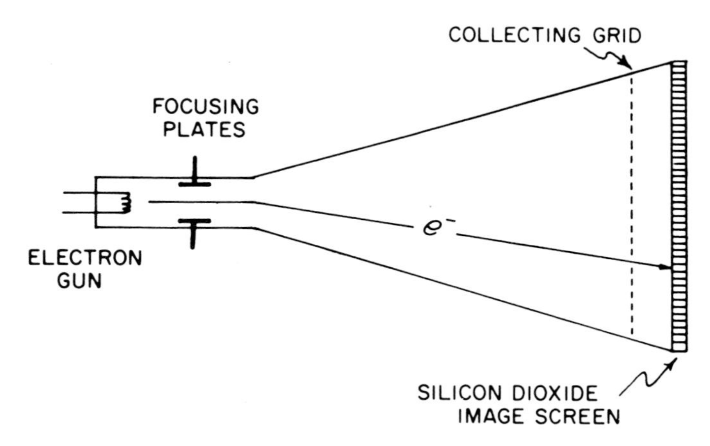

Figure 2. Image tube of an analog scan converter. Reproduced from *Abdominal Gray Scale Ultrasonography,* B. B. Goldberg (ed), by permission of John Wiley & Sons, New York, 1977.

any moment in time. This results in the development of a two-dimensional image on the surface of the scan converter tube. The image is an electronic image or virtual image which cannot be viewed directly by the operator. When the scan converter tube is switched from the "write" mode into the "read" mode, an electron stream is swept across the surface of the silicon dioxide image screen in a 525-line interlacing raster pattern identical to that of commercial television. Depending on the charge accumulated on the image screen during the "write" mode, the electrons are either attracted or repelled and collected on the collecting grid. The charges collected on the collecting grid are then amplified and reproduced on a TV monitor tube which can then be observed directly by the scan operator or photographed for permanent storage. Two disadvantages of these analog scan converter units are that they have a tendency to drift, requiring rather frequent servicing and adjustment, and that, although the speed of image writing is considerably faster than with the older bistable storage scopes, the "write" speed is still too slow to capture rapid scan motions such as might occur when one attempts to stop-frame a real-time ultrasound image or to capture rapidly moving structures such as the cardiac valves.

Recently, computer technology has developed to the point that digital storage devices have become increasingly less expensive and more widely used. Because of this, it is not surprising to discover that the ultrasound manufacturers are now replacing the analog scan converter units with digital scan converters. Similar to the storage units in computed tomographic scanners, digital scan converters consist of two-dimensional matrix of digital storage elements. These digital matrices are approximately 512 × 512 storage elements, allowing for approximately the same degree of resolution obtained with the analog scan converters. Digital storage units, however, have similar operating characteristics to the bistable storage scopes in that any given element is either on or off. A single plane of storage elements in a digital scan converter would then be expected to produce scans of the same restricted dynamic range as the older bistable storage scopes. However, by combining several layers or tiers together, it is possible to have more than one storage element in any given location. Depending on the number of tiers employed, it is possible to obtain scans in multiple shades of gray. Com-

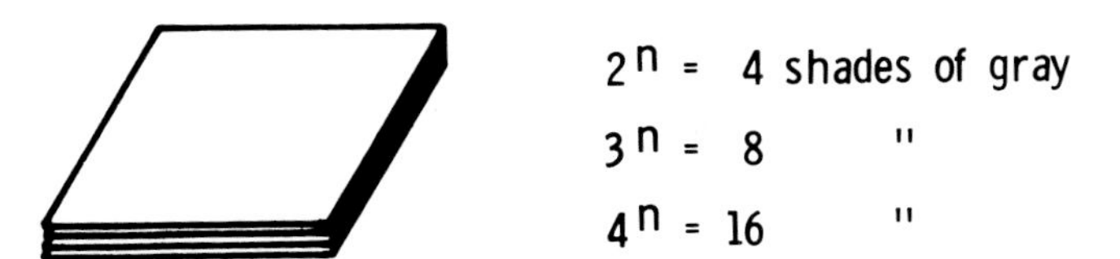

Figure 3. Schematic diagrams of a digital scan converter demonstrating a four-tiered memory plane. The numbers raised to the exponent "n" indicate the number of tiers of digital memory: "n" equals 2 for any digital memory unit.

mercially available scan converter units are now being produced with 4 or 5 image tiers allowing for 16 to 32 shades of gray (Fig. 3).

Digital scan converters permit extremely rapid writing speeds, allowing for capture of the most rapidly moving events, including stop-frame imaging of real-time scans. Digital scan converters also have the advantage over analog scan converters of being drift free. Once in the digital mode, it is possible to use already available digital hardware and software developments for processing the ultrasound signal in a variety of ways. Further clinical experience with digital scan converters will be necessary, however, before it can be concluded with any certainty whether the digital scan converters improve results in the clinical setting.

Mechanical Scanning Arms

Early investigators in ultrasound in the mid-1960s were aware of the fact that, for scan consistency and reproducibility, it was necessary to automate the action of the scanning arm. Automated mechanical arms were developed, but this required immersion of the patient in a water bath in order to achieve an adequate acoustical coupling of the transducer to the patient. Because this method was cumbersome, immersion scanners were abandoned in favor of mechanical scanning arms which could be directly applied to the patient without requiring emersion or the use of an intervening water bag. In use, this type of scanning arm requires considerable operator skill, and the scans obtained are highly dependent on the training and expertise of the technologist performing the study. Recently, a unit has been developed by Kossoff and others at the Commonwealth Acoustic Laboratory in Australia called the Octoson in which eight automatically driven transducers are submerged in a water bath. The patient lies on a thin plastic membrane floating on the surface of the water bath (Fig. 4). The technologist operating the machine is able to control the angle and direction of the transducer array, but the actual scanning motion of the transducers is independent of the technologist, resulting in consistent and reproducible scans free of operator-produced scanning artifacts. The automated motion of the scanner also permits rapid build-up of the scan with the required time to complete a single scan approximating 1 second.

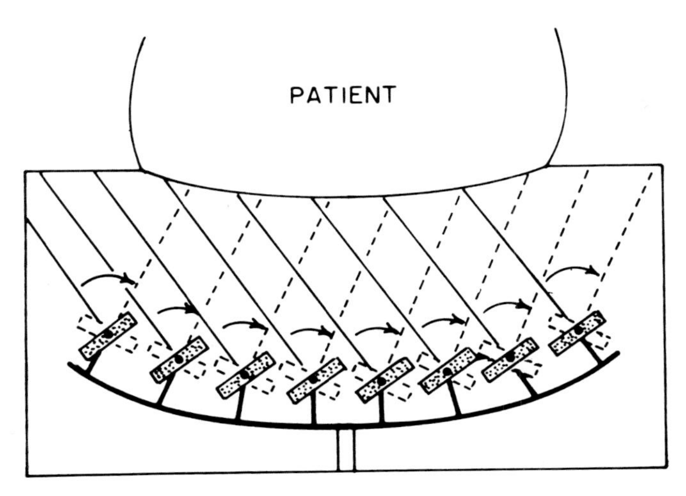

Figure 4. Schematic diagram of an automated mechanical scanning unit–the Octoson. Reproduced from *Abdominal Gray Scale Ultrasonography,* B. B. Goldberg (ed), by permission of John Wiley & Sons, New York, 1977.

Real-Time Scanners

Real-time scanners can be considered to be units that produce a series of static scans at a rate in excess of the fusion rate of the human eye, which is approximately 14 to 16 frames per second. The analogy would be a motion picture in which a series of static images pass in front of the eye above the visual fusion rate. Real-time scanners are available in two broad categories. One category of real-time scanners mechanically moves one or more transducers rapidly over the patient such that 14 or more scans are obtained per second. This gross type of mechanical transducer motion usually requires physical separation of the transducer from the patient by some type of water bath. Figure 5 is an example of this type of scanning unit. Another approach to real-time scanning has been to combine multiple transducers in a linear array with the transducers electronically switched such that the transducers are sequentially fired at a rate in excess of 14 frames per second again producing real-time images (Fig. 6). Complex firing patterns have been developed for these electronically switched linear array real-time units called multiplexing, phased array, and dynamic focusing, which offer certain advantages depending upon the intended use of the unit. Details concerning the relative advantages of these various modes of operation of real-time units as well as their diagnostic application are given else-

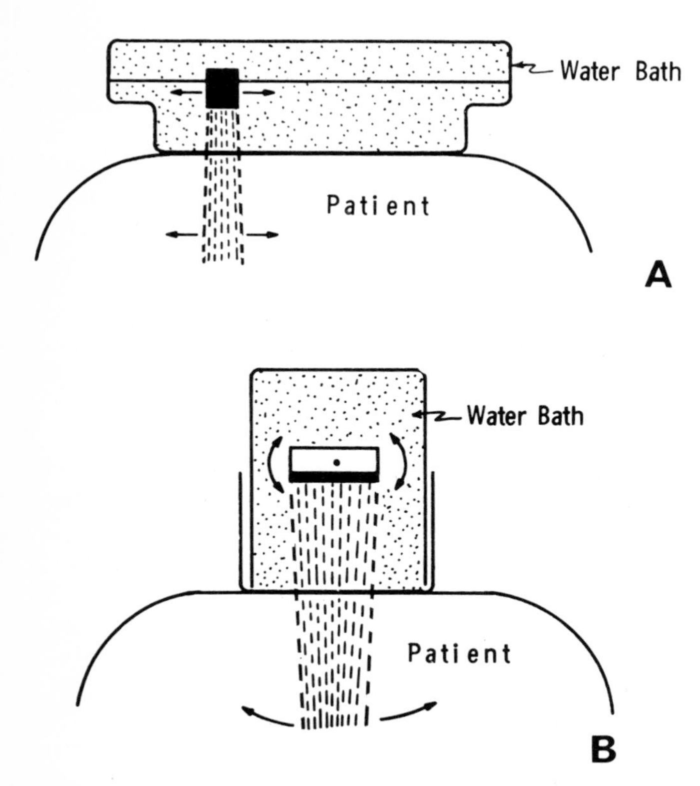

Figure 5. Schematic diagrams of two differing types of single transducer mechanical real-time units.

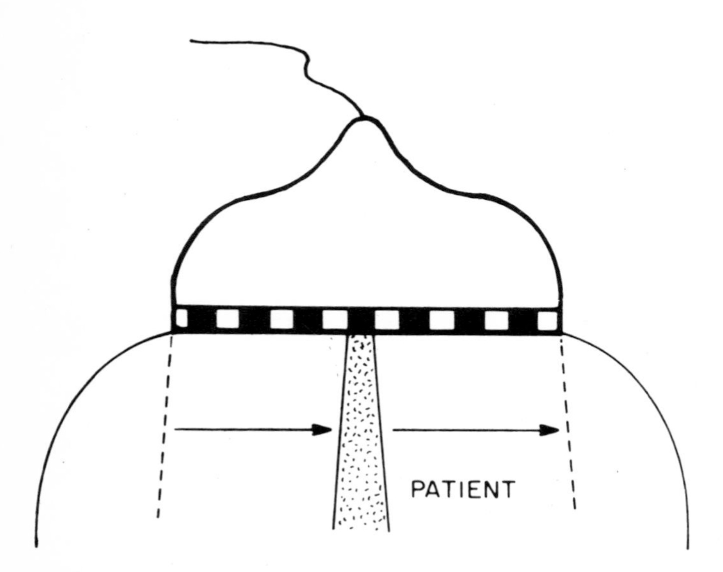

Figure 6. Example of an electronically switched multiple transducer array real-time unit. Reproduced from *Abdominal Gray Scale Ultrasonography,* B. B. Goldberg (ed), by permission of John Wiley & Sons, New York, 1977.

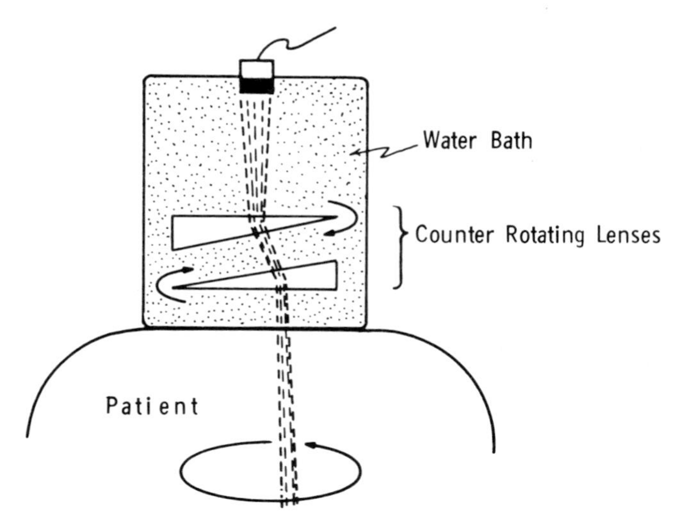

Figure 7. Schematic diagram of "C" mode real-time unit.

where in the literature and are beyond the scope of this presentation.[1]

A recent development in the field of real-time imaging is a unit that is capable of producing scans of varying depths in a plane paralleling the surface to which the transducer is applied. This plane of section, called the "C" mode, is perpendicular to the cross-sectional plane obtained with the more conventional B mode ultrasound units and is similar to that obtained with x-ray tomography. This type of ultrasound unit is still in the developmental stage and is not commercially available at this time, but does represent a technological breakthrough and will probably become commercially available in the not-too-distant future (Fig. 7).

Endoscanning

Although intracavitary use of ultrasound transducers has been available for some time, use has been limited and has not received broad clinical acceptance. Transducers have been designed for application to the tip of an examining finger for intracavitary use. Transducers of sufficiently small size have also been developed and applied to tips of catheters for use within blood vessels and in the heart. Single plane A/M mode scans are obtainable from these types of transducers. In addition, an intracavitary tube has been developed with an axial rotating transducer in its core for intracavitary use which provides 360-degree rotational B mode scans (Fig. 8). This type of scanning unit has been used transvaginally and transrectally for the evaluation of pelvic masses. Although this type of scanner has received only limited use to date, other more sophisticated endoscan-

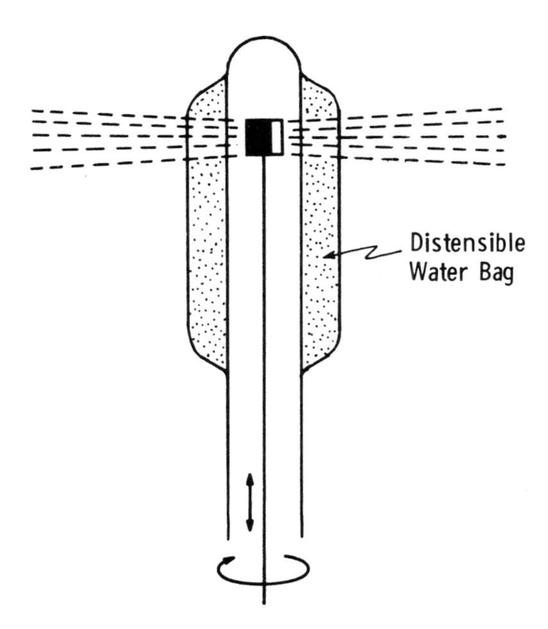

Figure 8. Example of a 360-degree rotating B mode endoscanner.

ning units currently under development at several centers can be anticipated to expand both the spectrum of application and the use of ultrasound in the area of endoscanning in the near future.

Summary

In summary, we have briefly surveyed the field of ultrasound, commenting upon recent or soon to be expected technological advances in ultrasound instrumentation that have directly affected or can be anticipated to affect the extent and quality of diagnostic ultrasound in the clinical setting.

References

1. Carlsen EN, Perlmutter GS: Gray Scale Instrumentation. In Goldberg BB (ed): Abdominal Gray Scale Ultrasonography. New York, Wiley, 1977, p. 12
2. Perlmutter GS, Carlsen EN: Physics of Gray Scale. In Goldberg BB (ed): Abdominal Gray Scale Ultrasonography. New York, Wiley, 1977, p. 1

Screening for Brain Disease

Fred S. Mishkin

In spite of vociferous advocates, there is not now, nor is there likely to be, a single, 100 percent precise screening procedure for brain disease. The clinician must, therefore, pick and choose between the multiple modalities available to confirm his opinion, the ones that will most likely give the correct answer with the least risk to the patient and with the most efficient use of the patient's (or third party payer's) funds. Ultimately, the physician must bear the burden for optimizing the cost/benefit ratio or it will be done by governmental agencies concerned more with the cost and social benefit than the benefit to the individual, since a bureaucracy has no sense of individual suffering.

There is no room for chauvinism in such a determination. The physician must determine what brain disease is most probably present and choose the most appropriate and efficient means of study. A rote routine is no substitute for a reasoned approach. The physician should always consider how the results of a particular study will influence the course of action, whether it will require confirmation by another technique, and, more particularly, how it will benefit the patient.

Computerized axial tomography provides superb anatomic detail with spatial and contrast resolution of brain tissue rivaled by no other noninvasive means. Use of iodinated contrast medium extends this anatomic detail to furnishing information concerning lesion vascularity. In detecting porencephaly, hydrocephalus, brain atrophy, benign cysts, and certain developmental abnormalities, the CT procedure stands alone. Radionuclide studies of the brain, although often approached from the anatomic point of view, in reality provide physiologic data supplemented by spatial orientation. Perfusion studies offer information concerning cerebral blood flow unavailable by other noninvasive means. The delayed radionuclide brain images offer information concerning the blood brain barrier.

Present data suggest that in detection of intracranial neoplasms both techniques *can*, in the appropriate hands, detect most neoplastic diseases of the brain with a similar precision. In nonneoplastic diseases, CT scanning appears much more sensitive in intracranial hemorrhage and acute trauma and is probably more sensitive in detecting demyelinating disease. The routine use of both procedures hardly seems justifiable in terms of patient benefit and increased cost. The physician, after considering local facilities and expertise, should select the study which is most likely to give a correct answer without further need for confirmation.

Specific Studies

Physicians have, in the past, chosen screening studies chiefly on the basis of their accuracy. The medical literature is replete with these kinds of studies for many techniques. Now public and, more stridently, governmental outcry has brought to the medical profession awareness of financial constraints. This necessitates that we justify retaining a screening technique on the basis of how well it manages to fulfill the considerations of cost, affect on outcome, patient benefit, and therapeutic possibilities. Such studies are beginning to appear.[1]

What is expected of a screening study for brain disease? First, it should be accurate, indicating the presence of disease only when it is present and not when it is not, as well as correctly indicating its absence. A screening procedure should, in fact, be weighted toward the false positive side so as not to give any false negative results and allow a potentially treatable lesion to slip by. Many physicians adhere wholesale to this principle, preferring to overinterpret so as not to run the risk of missing anything. This philosophy is increasingly burdensome today since it may lead to more costly, occasionally invasive procedures, sometimes with a morbid outcome for the patient. In this regard, it is pertinent to note that when physicians in Southern California withheld all but emergency services in response to the malpractice crisis in January 1976, the mortality rates in Los Angeles County, when compared with the average of the same weeks over the previous five years, showed a statistically significant decline after the first week of the slowdown, which persisted for the subsequent six weeks.

The screening study should involve little or no trauma to the patient and should be without risk. Ideally, the outcome of the screening test should alter the diagnostic and therapeutic plan. There is little sense in screening for a disease already virtually certain to be present unless the test provides pertinent information for therapy or concerning prognosis. The prognostic information may be valuable to the physician or to the patient or to both—particularly in terms of peace of mind in knowing that a disease is absent. Fortunately, most patients are not hypochondriacs with fear of a brain tumor. Unfortunately, most physicians have been indoctrinated with the necessity for compulsively excluding the worst and most terrifying diagnostic possibilities. Without available technology, this attitude was benign, although it gave the physician a great deal to worry about. Now, with the technology available, this attitude is proving very expensive. To screen all patients with symptoms such as headaches or seizures, which are unlikely to have a demonstrable organic cause, would be unrewarding in terms of yield and intolerably expensive. The screening test should be performed not to put our own minds at ease or reduce our malpractice risk, but rather because the result of the screening test will lead to improved patient welfare. Improved patient outcome or public health benefit is the only justification for performing a diagnostic study.

The diagnostic accuracy of a test is computed by establishing the relationship of the test outcome to the presence or absence of disease. If $T+$ indicates a positive test and $T-$ a negative outcome, and $D+$ indicates the presence of a disease whereas $D-$ indicates its absence, a simple decision matrix offers four possibilities. Of these, two are often used to assess the value of a test. First, the true positive value; that is, given the presence of a disease, the probability of a positive test outcome. True positive $= P(T+ \mid D+)$ indicates the sensitivity of a test. The other probability, the true-negative value, is, given the absence of disease, a negative test outcome. True negative $= P(T- \mid D-)$ represents the specificity of the test. The sum of these correct outcomes divided by all outcomes gives the accuracy of the test.[2] The physician, in studying an individual patient, needs to know the inverse of these probabilities. For example, given a positive test result, what is the probability that a disease is present? And given a negative test result, what is the probability that a disease is absent?

How do the two most valuable tests for brain disease compare as screening tests? The radionuclide study is essentially free from all except radiation hazards and has the minimal discomfort of a venipuncture and immobilization for positioning. The CT study, while in itself noninvasive, is often routinely accompanied by injections of contrast medium, a procedure that has a small but certain number of severe reactions, occurring in approximately one patient in ten thousand. More pertinent are the consequences of more frequently evoked nausea and vomiting in a patient who must be immobilized in the supine position, setting the stage for aspiration, particularly in the semicomatose individual. While some motion is tolerable on the radionuclide study, which has only gross resolution at best, any motion intolerably degrades the CT study. Thus, sedation and anesthesia, as well as their attendant risks, will more often be needed with CT scanning. In terms of radiation risk, the CT irradiation, while significant, is at least with appropriate collimation confined to the head, whereas with radionuclide studies the entire body is radiated. Thus, a radionuclide study during pregnancy would irradiate the fetus, whereas the CT study would not.

While the radionuclide study has high sensitivity for a number of conditions, some it misses altogether. Many of these are conditions which are readily detected by the CT scan. It is, however, incorrect to believe that all brain abnormalities are accompanied by detectable changes in brain density.[3] Some diseases cause no density changes, or ones that are too subtle to discriminate. At other times, density changes will not be demonstrated due to technical factors such as not including the abnormality on one of the CT slices, or motion obscuring the lesion. In spite of the fact of occasional rare misses, the *overall* accuracy of the CT scan is undoubtedly superior to the radionuclide study in most, if not all, categories of brain disease. Some have suggested an increased accuracy may be obtained by performing both studies,[4] but such a policy as a routine has too little increased yield to justify the additional expense.[5] It is better for individual patients' welfare to understand what each technique does and what it does well so that the appropriate technique may be chosen to yield the most useful results most efficiently rather than follow a routine.

What is depicted by each study? The radionuclide study shows the anatomy of the brain only indirectly by outlining the vascular structures of the head and brain. On the other hand, the CT scan offers an unrivaled, detailed view of the brain itself differentiating white from gray matter as well as delineating normal cisterns, sulci, and ventricular spaces with anatomic textbook clarity. None of this can be done by the radionuclide study. Both studies, particularly the CT scan after contrast administration, can outline the large vascular pools surrounding the brain, but the CT study has the advantage of actually being able to depict the cross section of the vessel itself. Both studies can delineate a blood brain barrier deficit, the CT scan by leaking contrast medium into the lesion and "enhancing" the lesion density and detectability. Cerebral perfusion can be estimated very simply and quickly in a physiologic sense by the radionuclide transit study, whereas the CT scan does this only in an anatomic fashion. With this background, what follows is an attempt to evaluate the use of each of these modalities as a screening procedure in a variety of common brain diseases.

CONGENITAL DISEASES The CT images, by delineating the major intracranial structures such as cisterns, ventricles, falx, and sulci, can detect congenital lesions such as obstructive hydrocephalus, cystic malformations, and atrophy consequent to a variety of processes.[6] The radionuclide study detects only the most gross of such lesions, and then adds little more than is apparent on the skull film. With regard to AV malformations, the sensitivity of the radionuclide perfusion study is quite good, missing only the smallest of such lesions.[7] In some cases, it may detect lesions not seen on the CT scan, particularly if the CT study is done without employing contrast medium enhancement. Angiography, of course, is best suited to detect arteriovenous malformations and probably it is the only true screening test for such potentially curable lesions.

INFLAMMATORY LESIONS In acute diffuse meningitis and cerebritis, neither the CT scan nor the radionuclide study offers much of value.[8] In the more aggressive encephalitides such as herpes simplex encephalitis, the tissue density changes will be apparent on the CT scan and the blood barrier disruption will be demonstrable on the radionuclide study. The more indolent proliferative changes of pachymeningitis produced by tuberculosis or chronic *H. Influenza* meningitis may produce detectable anatomic changes demonstrable by CT scanning and may be accompanied by vascular alterations visible on the radionuclide study. The radionuclide study and CT scan are most useful in detecting localized inflammatory processes such as abscess or empyema.[8,9] Both these screening techniques may, in fact, be more sensitive than angiography early in the course of these brain diseases and there is little to choose between the two other than which is most readily available.

TRAUMA Radionuclide studies are poorly suited for the study of the acute effects of trauma since clinically insignificant, superficial injuries such as scalp contusions have significant tracer uptake often indistinguishable from intracranial lesions. In contrast, CT scanning can readily demonstrate the acute effects of tissue density change and brain displacement resulting from trauma.[10] On the other hand, as the lesion evolves into the more chronic state, the radionuclide study becomes more useful, and at times the changes on CT scan may actually become much more subtle. Figure 1 is an example of bilateral subdural hematomas which have become isodense with normal brain tissue. Although the skilled interpreter made the correct diagnosis, those less well trained did not, in spite of the fact the diagnosis was suspected by the referring physician. In this case, the accompanying radionuclide study provided supporting data so that angiography was performed to confirm the large subdurals, which extended into the prefrontal area. Thus, while the CT study is the first line of approach in the traumatized patient, radionuclide imaging may add crucial information in selected cases.

CEREBROVASCULAR DISEASES Infarcts appear as lesions of varying density on the CT scan with variable detectability. Many are visible during the acute stage due to surrounding edema and evolve into chronic lesions of low density.[11] Radionuclide studies similarly have an evolutionary pattern infarction.[12] They are abnormal at some time during the course of approximately two-thirds of patients who have completed stroke, although usually not during the first few days. The radionuclide abnormality usually resolves in two months, even though the neurologic deficit persists. The abnormality seen on radionuclide imaging seems to correspond with the ability of the lesion to enhance with contrast on CT scan.[13] Overall, the CT scan is more sensitive in detecting acute infarction and the chronic sequelae of tissue destruction.[14] The radionuclide perfusion study may be helpful in offering prognostic information. We found during the acute episode that if the venous phase of the radionuclide perfusion study showed increased activity on the ischemic side as compared with the normal hemisphere, the patient had a better chance for making some functional recovery.[15] This finding is presumably related to adequacy of collateral circulation. In about half the patients studied acutely with transient ischemic attacks, the radionuclide angiogram will show diminished perfusion to the affected hemisphere, whereas the CT scan will be normal. However, the CT scan has shown that some patients suspected of having only ischemia have actually had infarcts. Neither study offers the detailed vascular anatomic information prerequisite for surgical therapy for correction of the ischemia underlying the infarct.

Although some have found radionuclide studies useful in detecting intracerebral hematomas,[16] our experience has been poor in detecting them in a population with a high incidence of intracerebral bleeding accompanying hypertensive cardiovascular disease. The CT scan detects and accurately localizes intracerebral hematomas with a precision which can often predict the etiology.[17] This information can be helpful in deciding upon surgical therapeutic intervention. The benefit of this to the patient remains to be demonstrated.

One advantage of the radionuclide study is the ability to diagnose cerebral death by documenting cessation of cerebral perfusion.[18] This use demonstrates the physi-

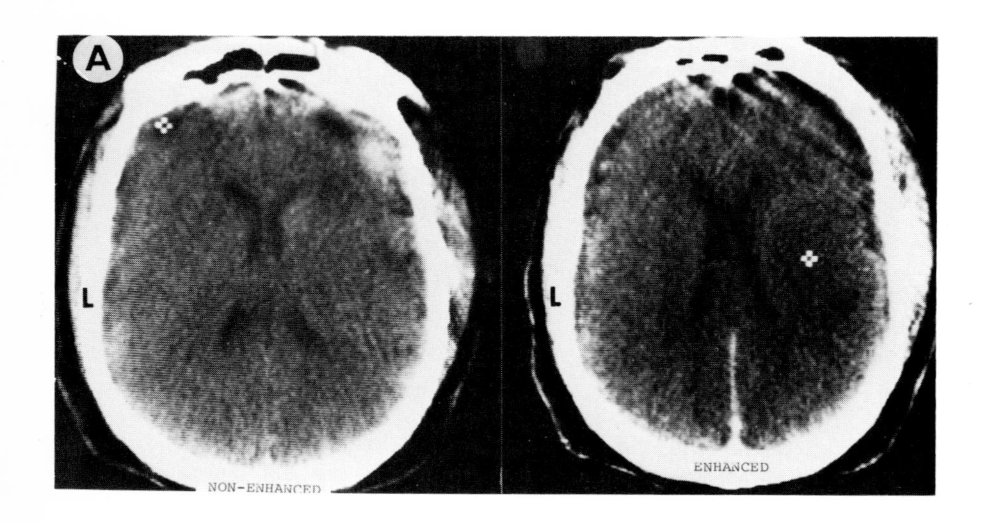

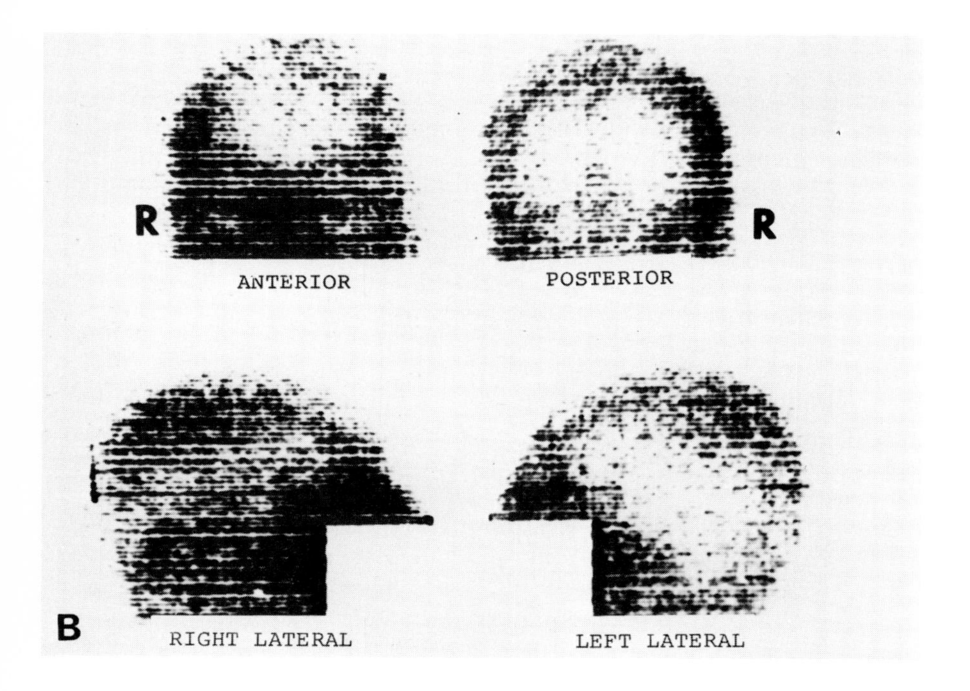

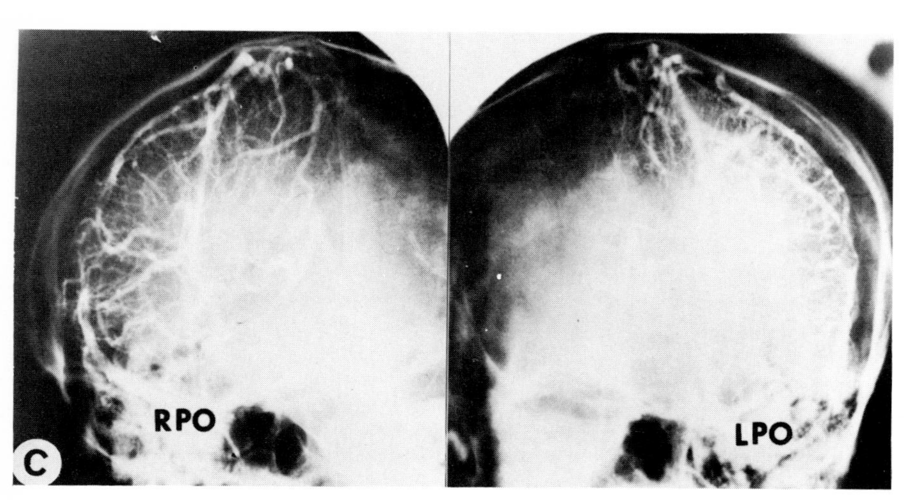

Figure 1. Bilateral chronic subdural hematomas (shown in Parts A, B, and C). **A.** The unenhanced CT study shows density increase over the periphery of the right frontal hemisphere. No brain displacement is evident. The enhanced CT study shows that the superficial prefrontal areas bilaterally fail to increase in density since they contain hematomas rather than normally perfused brain. **B.** The radionuclide images show increased activity over both convexities, particularly on the right. The left lateral view shows thickening at the vertex. **C.** The oblique angiograms show displacement of the cerebral vascularity away from the skull vault bilaterally. Sizable bilateral frontal and prefrontal subdural hematomas were surgically evacuated.

ologic approach that is unique to the radionuclide methodology.

NEOPLASMS Both radionuclide scanning and computerized tomography are remarkably sensitive means for detecting intracranial neoplasms. This poses a dilemma. Because of its exquisite anatomic delineation and ability to detect smaller lesions, including the effect of the lesion on surrounding structures, as well as its greater overall sensitivity, notably in low grade neoplasms, the CT scan is unquestionably the screening procedure of choice.[19,20] If this is the case, should radionuclide brain imaging for neoplasms be scrapped? Probably not, since it has very high sensitivity, particularly in detection of the meningioma and glioblastoma; but it is no longer the first line of approach, nor should it be used routinely in addition to the CT study. Tumors in certain areas, such as the base of the skull, pituitary fossa and brainstem, are nearly always inaccessible to detection by routine radionuclide studies. The radionuclide brain scan should not be used if lesions in these areas are being sought. Should the radionuclide study be done with the knowledge that it will have to be followed by a CT study? No, it is too expensive. Thus, if one is searching for metastasis and the discovery of even a single metastasis is sufficient to alter the therapeutic approach, then the radionuclide study may be sufficient. If more accurate information is needed, for example, if resection is planned, then the radionuclide study should be bypassed in favor of the CT scan. Only if the CT scan does not seem to provide sufficient information for decision making should the radionuclide study be used in addition to the CT study.

Posterior fossa lesions seem to be more difficult to detect for both the screening procedures. Initial results suggest about equal sensitivity[21]; however, given the basic detection principles of these procedures, larger series will probably demonstrate the superiority of the CT scan.

MISCELLANEOUS DISEASES Demyelinating diseases may cause abnormalities on the radionuclide study during their acute phases,[22] but they do not do so with great regularity. The acute phases of some of these diseases may cause detectable density differences on the CT study, but the sensitivity remains to be determined. The chronic sequelae of the degenerative diseases, particularly atrophy, are most readily detectable by CT imaging.[23] In patients with such suspected processes, the CT scan is the first choice in screening. As in the child, CT scanning in the adult sensitively detects changes in ventricular size and is the method of choice for case finding in suspected hydrocephalus. Other methods, such as pneumoencephalography or radionuclide cisternography, may be necessary to complete the diagnostic information in order to make a therapeutic choice.

Table 1. Which tests for screening?

	RN	*CT*
Congenital		+
AVM	+	
Inflammatory	Either	
Trauma		
Acute		+
Chronic	*	+
Cerebral vascular disease	*	+
Neoplasms	*	+
Hydrocephalus		+

(+) Procedure of choice; (*) RN is second line of attack if CT is normal. Routine use of both is unwarranted.

Conclusion

Which tests should be used for screening brain disease? Not both. Table 1 gives a capsule summary of the opinions presented in this overview. CT scanning, except in inflammatory disease where either technique appears as effective, is by and large the procedure of choice. In the search for congenital arteriovenous malformations, the radionuclide technique may be more sensitive but, in general, congenital abnormalities are better detected by CT scanning. Similar conclusions apply to traumatic lesions, cerebral vascular disease, and neoplasms, leaving radionuclide studies the role of a back-up screening procedure. There are enough exceptions and individual idiosyncracies that well executed rationuclide brain imaging is still useful. A poorly done and interpreted study with either technique has little value to the patient, no matter how good published results appear. Often what is chosen for a particular patient depends upon available local facilities and interest. Thus, the radionuclide brain scan in the hands of an interested expert can provide more valuable information than a CT study performed casually by an uninterested individual. As CT scanning becomes more and more widely used by less interested and less well-motivated individuals, its accuracy as reported by large research centers will drop somewhat. Furthermore, it will suffer the fate of all newly introduced, popularized screening tests. Since the prevalence of a disease remains unchanged, many more individuals will be screened to find nearly the same number of abnormal studies. Given fixed capital and operational expenditures, the cost per study will decrease, but the cost of finding a single case of disease will increase.

References

1. McNeil BJ, Adelstein SJ: Measures of Clinical Efficacy. The Value of Case Finding in Hypertensive Renovascular Disease, N Engl J Med 293:221, 1975

2. McNeil BJ, Keeler E, Adelstein SJ: Primer on Certain Elements of Medical Decision Making, N Engl J Med 293:211, 1975
3. Fordham EW: Complementary Role of Computerized Axial Transmission Tomography and Radionuclide Imaging of the Brain, Sem Nucl Med 7:137, 1977
4. Christie JH, Mori H, Go RT, Cornell SH, Schapiro RL: Computerized Tomography and Radionuclide Studies in the Diagnosis of Intracranial Disease, Am J Roentgenol 127:171, 1976
5. Evens RG, Jost RG: Clinical Efficacy and Cost Analysis of Cranial Computed Tomography and the Radionuclide Brain Scan, Sem Nucl Med 7:129, 1977
6. Harwood-Nash D: Congenital Craniocerebral Abnormalities and Computed Tomography, Sem Roentgenol 12:39, 1977
7. Rosenthall L: Radionuclide Diagnosis of Arteriovenous Malformations with Rapid Sequence Brain Scans, Radiology 91:1185, 1968
8. Claveria LE, du Boulay GH, Mosely IF: Intracranial Infections: Investigation by Computerized Axial Tomography, Neuroradiology 12:59, 1976
9. Crocker EF, McLaughlin AF, Morris JG, Benn R, McLeod JG, Allsop JL: Technetium Brain Scanning in the Diagnosis and Management of Cerebral Abscess, Am J Med 56:192, 1974
10. Ambrose J, Gooding MR, Uttley D: EMI Scan in Management of Head Injuries, Lancet 1:847, 1976
11. Gado MH, Coleman RE, Merlis AL, Alderson PO, Lee KS: Comparison of Computerized Tomography and Radionuclide Imaging in Stroke, Stroke 7:109, 1976
12. Verhas M, Schoutens A, Demol A, Patte M, Demeurisse G, Ganty CH, Rakovsky M: Study in Cerebrovascular Disease: Brain Scanning with Technetium 99m Pertechnetate. Clinical Correlations, Neurology 25:553, 1975
13. Masden JC, Azar-Kea B, Rubino FA: Evaluation of Recent Cerebral Infarction by Computerized Tomography, Arch Neurol 34:417, 1977
14. Chiu LC, Christie JH, Schapiro RL: Nuclide Imaging and Computed Tomography in Cerebral Vascular Disease, Sem Nucl Med 7:175, 1977
15. Barrett IR, Powell FD, Mishkin FS: Prognostic Value of Radionuclide Angiography in Cerebral Vascular Disease (Ab), J Nucl Med 17:527, 1976
16. Planiol TH, Degiovanni E, Groussia P, Ihi R, Gouaze A: Intracerebral Hematomas in Gamma-Angioencephalography, Rev Neurol 131:301, 1975
17. Hayward RD, O'Reilly GVA: Intracerebral Hemorrhage: Accuracy of Computerized Transverse Axial Scanning in Predicting the Underlying Etiology, Lancet 1:1, 1976
18. Mishkin FS: Cerebral Radionuclide Angiography, Angiology 28:261, 1977
19. Mori H, Lu CH, Chiu LC, Cancillia PA, Christie JH: Reliability of Computed Tomography: Correlation with Neuropathologic Findings, Am J Roentgenol 128:795, 1977
20. Alderson PO, Gado MH, Siegel BA: Computerized Cranial Tomography and Radionuclide Imaging in the Detection of Intracranial Mass Lesions, Sem Nucl Med 7:161, 1977
21. Mikael MA, Mattar AG: Sensitivity of Radionuclide Brain Imaging and Computerized Transaxial Tomography in Detecting Tumors of the Posterior Fossa, J Nucl Med 18:26, 1977
22. Gize RW, Mishkin FS: Brain Scans in Multiple Sclerosis, Radiology 97:297, 1970
23. Huckman MS, Fox JH, Ramsey RG: Computed Tomography in the Diagnosis of Degenerative Diseases of the Brain, Sem Roentgenol 12:63, 1977

The Comparative Results of Brain Pathology: RN Versus CT

Lee C. Chiu
Victoria S. Yiu
Rolf L. Schapiro

Radionuclide brain scanning (RN) and cranial computed tomography (CT) are the most important noninvasive diagnostic techniques applied to the evaluation of the patient with suspected intracranial disease. The former enjoys a long history of clinical experience, and the rapid evolution of CT into a neurodiagnostic technique renders continued comparison of the relative effectiveness of these diagnostic modalities desirable.

This article represents a comparative review of 854 patients with a variety of brain disorders who were evaluated with both RN and CT scans at The University of Iowa Hospitals and Clinics. It is observed that in some disease states, one modality is clearly superior to the other, permitting delineation of the most efficacious technique. In other disorders, both modalities provide complementary information that is important in defining the most appropriate therapy.

Materials and Methods

Between September 1973 and January 1977, more than 6000 CT scans and 7500 RN brain scans were performed at The University of Iowa Hospitals and Clinics. The study consisted of 854 patients who had both CT and RN scans at 1- to 3-week intervals. Both diagnostic modalities were scheduled and performed independently of one another as requested by the referring physician and no attempt was made to influence the clinician's choice of initial study to be performed.

The RN scans were performed on either a Nuclear Chicago Photo-gamma III Camera or a Picker Dyna DC4 Camera. All patients were given 400 mg KCLO4 at least 20 minutes prior to the administration of the radionuclide. After bolus injection of 25–30 mCi [^{99m}Tc] pertechnetate through the antecubital vein, a rapid sequential flow study was routinely obtained in the anterior view at 2-second intervals for a total of 32 seconds. Static brain images were obtained in anterior, posterior, and both lateral views at variable time intervals ranging from 10 minutes to 2 hours after the injection of radionuclide. Additional views (vertex view) were obtained as deemed necessary. If the clinical history suggested the possibility of neoplastic disease, CVA, brain abscess, or subdural hematoma, 3- to 4-hour delayed static scans were also obtained.

The CT scans were done on an EMI Mark I, EMI CT1000 scanner or on an EMI CT5005 scanner. Eight tomographic slices were routinely taken at different levels depending on the area of interest. Studies with contrast agents were performed selectively as based on the initial scans or as deemed desirable on the basis of the patient's clinical history. Approximately one-third of the entire patient population was scanned both before and after the infusion of contrast material.

Results

There were 340 patients with primary or secondary, malignant or benign intracranial neoplasms. CT detected 90 percent of these lesions, RN, 82 percent, and the combination of both studies detected 93 percent of these tumors; the overall, known false negative rate of the combination of the two screening procedures is therefore limited to 7 percent. Of these 340 patients, 237 had lesions in the supratentorial regions; in this location the detection rate for CT was 92 percent, for RN 87 percent, and the combination of both screening procedures identified 94 percent of the lesions. In 41

Table 1. Result of dual CT/RN studies (+) (−) for neoplasm

Location	*No.*	*CT+ RN+*	*CT− RN−*	*CT− RN+*	*CT+ RN−*	*Total (%) CT+*	*RN+*	*Both*
Supratentorial	237	201	14 (6%)	5	17	92	87	94
Base of brain	41	22	5 (12%)	2	12	83	59	88
Posterior fossa	62	48	6 (10%)	1	7	89	79	90
Total	340	271 (80%)	25 (7%)	8 (2%)	36 (11%)	90	82	93

of the 340 patients, the tumor was located at the base of the brain. In this location, CT detected 83 percent of the lesions, RN only 59 percent, and both techniques combined had a diagnostic yield of 88 percent. In 62 patients the lesions were located in the posterior fossa; CT examinations were positive in 89 percent of these, RN scans were positive in 79 percent, and the combined yield of the two techniques was 90 percent. Table 1 summarizes the results of CT and RN scans in patients with neoplastic disease.

There were 174 patients with occlusive disease of the middle cerebral artery (MCA) and relatively recent infarction. The main MCA was involved in 117 of these patients. Of these 11 were examined within 1 week of the onset of clinical symptoms; CT was positive in 80 percent and RN in 68 percent of the cases, and the combined positive yield was 82 percent. From 2 to 4 weeks after the neurologic deficit became apparent, 106 patients were examined. Within this time frame, CT was positive in 82 percent and RN in 88 percent of the cases, and the combined positive yield was 88 percent. Cortical branches of the MCA appeared to be involved in 46 of the 174 patients. CT examinations were positive in 70 percent, while RN examinations were positive in 86 percent of these patients, and the combination of these examinations yielded a positive diagnosis in 87 percent of patients. Infarcts in the region supplied by the deep branch of the MCA were identified in 11 of the 174 patients. CT picked up 9 of these cases and RN demonstrated the lesion in 8 patients. Recent infarcts in the posterior fossa or attributable to occlusive disease of the posterior cerebral artery were detected by both CT and RN in 80 to 90 percent of cases. Demonstration of scars attributable to remote infarcts is achieved in a great proportion (90 percent) of CT scans, but only rarely (4 percent) on RN scans. The data on comparative sensitivity of CT and RN scans in regard to occlusive vascular disease are summarized in Table 2.

There were 98 cases of intracerebral hemorrhage due to causes other than trauma, infarction, or neoplasm. CT scans were positive in 94 percent of these cases, while positive RN scans were limited to 48 percent of cases. Vascular anomalies such as arteriovenous malformations (AVM) or aneurysms, without associated acute hemorrhage were present in 48 patients and detected by CT in 87 percent, and by RN scan in 72 percent of cases. Aneurysms and AVM detected by CT were generally greater than 1.5 cm in diameter and most of the patients were examined with contrast-enhanced CT scans.

Patients with intracranial infections included 38 cases of brain abscesses, either single or multiple. CT successfully demonstrated these in 92 percent of cases and RN scans in 82 percent of cases. The combined diag-

Table 2. Comparative sensitivity of RN and CT scans in CVA (total 272 cases)

	Total (%) CT	RN	Both
I. Middle cerebral artery infarction (1 to 4 weeks)			
A. Main MCA Branch (117)			
Less than one week (11)	80	68*	82
2nd to 4th week (106)	82	88	88
B. Cortical branch (46)	70	86	87
C. Deep branch (11)	78	72	78
Total (174)	74	79	84
II. Posterior cerebral artery infarction (32)	84	88	90
III. Posterior fossa artery infarction (13)	80	78	84
IV. Old MCA or other infarction (53)	90	4	90

* Dynamic scan (+).

Table 3. Overall sensitivity of CT and RN scans

Type of lesion	Number of studies	Accuracy (%)		
		CT	RN	Combined
Neoplasm				
(primary, secondary malignant & benign tumors)	340	90	82	93
Occlusive vascular disease	272	74	79	84
Cerebral hemorrhage	98	94	48 (33)	96
Vascular abnormality				
(AVM aneurysm, etc.)	48	87	72	90
Infection:				
(Encephalitis)	13	81	77	86
(Abscess)	38	92	82	94
Head trauma (extracerebral collection)				
(subacute)	45	75	88	92
(chronic)	108			
	63 (CT)	85	—	—

nostic yield was 94 percent. Infectious encephalitis was encountered in 13 patients. CT shows nonspecific low density changes in 11 of these patients, and 10 patients showed abnormalities on RN scans. The latter, however, rendered a quite specific diagnosis of herpes encephalitis in 6 patients who showed a characteristic finding of localized, early radionuclide uptake in the temporal region that persisted for 2 or more hours.

There were 108 cases of extracerebral collections secondary to acute trauma, but only 45 of these were examined with both modalities. In this group, CT examinations were positive in 75 percent of cases, and RN scans were positive in 88 percent of cases, with an overall combined positive yield of 92 percent. Subsequent patients (63 cases) were screened with CT only prior to angiography and/or surgery, and in this group, CT rendered a positive diagnosis in 85 percent. The comparative data pertaining to cerebral hemorrhage, vascular abnormalities, infectious disorders, and acute trauma are included as part of Table 3.

Discussion

Neoplasm

The capability of CT to detect intracranial tumors exceeds that of RN scanning by about 8 percent. Since the resolving power of CT is considerably greater than that of scintigraphy, it appears that there are certain lesions that are more readily detected by CT scanning (e.g., low-grade cystic gliomas and multiplicity of metastatic lesions).[1,2] CT can detect smaller and presumably earlier tumors; the smallest lesion detected by CT in the present review measured 0.5 cm in diameter.[3] At times, a diagnosis is possible with even smaller lesions, provided the small neoplasm exhibits intense contrast enhancement, or if the tumor is surrounded by a substantial volume of edematous brain tissue. In the latter circumstance, the diagnosis will of course be nonspecific and may require clarification by invasive diagnostic maneuvers. Additional practical advantages of CT include the demonstration of cystic, hemorrhagic, and calcific components of the neoplasm and the ability to delineate secondary anatomic distortions in terms of ventricular size and shift, and displacement of the pineal gland and choroid plexus.[4,5] These manifestations secondary to the mass effect in many cases bear significantly on the planning of the most appropriate therapeutic approach and possibly the need for relatively early surgical intervention. On the other hand, nonenhanced CT scans may pose difficulty in differentiating neoplasm from the surrounding edema, and hence it may be easier to define the true size of the tumor by RN scanning. In patients with suspected intracranial neoplasm, our preferred initial examination is a CT scan; if the CT scan is positive, angiography is undertaken for confirmation and preoperative delineation of the tumor's vascular supply. If the CT scan is negative and there is some reluctance to subject the patient to cerebral arteriography, RN scans should be performed for further clarification before a final decision on the advisability of angiography is made.

Occlusive Vascular Disease

In cerebrovascular accidents with duration of clinical symptoms for less than 10 days the initial and followup examinations should be CT scanning. If the acute neurologic deficit became manifest more than 10 days prior to the examination, the preferred initial examination is either an RN or CT scan, and possible follow-up examinations by CT are helpful in determining the patient's progress by delineating the degree of hemorrhage and/or edema associated with the infarct.[6–10]

Intracranial Bleeding

Because of the clearcut superiority of CT over RN in the definition of intracerebral hemorrhages, a patient suspected of this catastrophic event on clinical grounds should be examined first with the CT technique. Extracerebral, intracranial hemorrhage secondary to trauma can be screened most reliably with RN studies. Yet, in our own clinical environment, CT has become the first and virtually exclusive screening procedure applied to the severely traumatized patient. There are several considerations that explain this apparent paradoxical contradiction. Although the RN flow study provides a quite reliable presumptive diagnosis of subdural hematoma, a further convincing and definitive diagnosis is usually deferred until 2 to 4 hours have passed and static brain scans are available as well. Probably more important is the fact that CT provides information on several other possible effects of trauma. These include cerebral contusion and hemorrhage, displacement or deformity of the ventricular system and/or the pineal gland reflecting any intracranial space-occupying lesion, and the presence of air within the subarachnoid space. Because of these considerations, the initial screening procedure for acute head trauma is CT in our clinical environment. Confirmation of a positive CT scan by angiography may or may not be interposed between the CT scan and craniotomy, a decision usually made by the neurosurgeon and based on his assessment of the urgency for operative intervention. If the patient's clinical condition permits a deliberately paced diagnostic workup, a negative CT scan should reasonably be followed by an RN or repeat CT scan for further clarification before a decision on more invasive and hazardous procedures is finalized.

Vascular Anomalies

We believe that a clinically suspected vascular abnormality should be evaluated initially with angiography. Radiologic screening procedures appear to have limited application in these disorders unless there are relative, but specific contraindications to angiography.

Intracranial Infections

Because quite specific characteristics permit in many instances a specific CT diagnosis of abscess[11,12] this modality should be the procedure of choice initially as well as in follow-up for response to therapy. The converse holds true for herpes simplex encephalitis and RN technique is the procedure of choice for these disorders.[13]

CT Versus RN

In conformity with the expressed opinions of others[14] there is little doubt that CT is the initial screening procedure of choice in the evaluation of the patient with suspected brain disease. This contention is supported by the empirical observation that in most common brain disorders, CT provides the greatest probability of detecting a focal lesion, has a reasonable chance of delineating a relatively narrow but meaningful differential diagnosis, and quite accurately projects the effects of a focal lesion on the remaining, intrinsically normal intracranial contents. Scintigraphic brain examinations primarily should be reserved for further clarification of those cases in which the CT diagnosis does not satisfactorily explain the patient's clinical presentation. Without doubt, in most disorders the validity of the combined results of complementary CT and RN examinations exceed that of the results of either examination alone. Finally, it must be remembered that the "routine" CT examination and its interpretation are not as demanding as an RN examination. There is little question that the success of an RN examination in approaching the diagnostic reliability of a CT examination is predicated not only on interpretative expertise, but also on the ability and willingness for meticulous performance of the examination which must be suited to the specific clinical problem posed by the individual patient. Four views of the calvarium, obtained 1 or 2 hours after injection of the radionuclide, are not likely to yield the diagnostic information comparable to the results reported in this review.

References

1. New PFJ, Scott WR, Schur JA et al.: Computed Tomography with the EMI Scanner in the Diagnosis of Primary and Metastatic Intracranial Neoplasm, Radiology 114:75, 1975
2. Passalaqua AM, Braunstein P, Kricheff II, et al.: Clinical Comparison of Radionuclide Brain Imaging and Computerized Transmission Tomography II—Noninvasive Brain Imaging: Computed Tomography and Radionuclides. Society of Nuclear Medicine 1975, pp. 173–181
3. Mori H, Lu CH, Chiu LC, Cancilla PA, Christie JH: Reliability of Computed Tomography: Correlation With Neuropathologic Findings, Am J Roentgenol, Radium Ther Nucl Med 128:795, 1977
4. Cornell SH, Christie JH, Chiu LC, Lyon LW: Comput-

erized Axial Tomography of the Cerebral Ventricles and Subarachnoid Spaces, Am J Roentgenol Radium Ther Nucl Med 124:186, 1975
5. Cornell SH, Musallam JJ, Chiu LC, Christie JH: Individualized Computer Tomography of the Skull With the EMI Scanner Using the 160 × 160 Matrix, Am J Roentgenol Radium Ther Nucl Med 126:779, 1976
6. Chiu LC, Fodor LB, Cornell SH, Christie JH: Computed Tomography and Brain Scintigraphy in Ischemic Stroke, Am J Roentgenol Radium Ther Nucl Med 127:481, 1976
7. Go, RT, Chiu LC, Neuman LA: Diagnosis of Superior Sagittal Sinus Thrombosis by Dynamic and Sequential Brain Scanning, Neurology 23:1,199, 1973
8. Chiu LC, Christie JH, Schapiro RL: Nuclide Imaging and Computed Tomography in Cerebral Vascular Disease, Semin Nucl Med 7:175, 1977
9. Yock DH Jr, Marshall WH Jr: Recent Ischemic Brain Infarcts at Computed Tomography: Appearance Pre- and Post-contrast Infusion, Radiology 117:599, 1975
10. Cronqvist S, Brismar J, Kjellin K, et al.: Computer Assisted Tomography in Cerebrovascular Lesions, Acta Radiol [Diagn] 16:135, 1975
11. Chiu LC, Jensen JC, Cornell SH, Christie JH: Computerized Tomography of Brain Abscess, Comp Axial Tomogr 1:33, 1977
12. Zimmerman RA, Patel S, Bilaniuk LT: Demonstration of Purulent Bacterial Intracranial Infections by Computed Tomography, Am J Roentgenol Radium Ther Nucl Med 127:155, 1976
13. Karlin CA, Robinson RG, Hinthorn DR, and Liu C: Radionuclide Imaging in Herpes Simplex Encephalitis, Radiology 126:181, 1978
14. Pendergrass HP, McKusick KA, New PFJ, Potsaid MS: Relative Efficacy of Radionuclide Imaging and Computed Tomography of the Brain, Radiology 116:363, 1975

Brain Imaging of Cerebrospinal Fluid: Computed Tomography and Nuclear Medicine Correlation

C. Leon Partain
Edward V. Staab

Cerebrospinal fluid imaging utilizing computed tomography (CT) and nuclear medicine are useful diagnostic tools in the evaluation of dementia; the most significant diagnostic dilemma being normal pressure hydrocephalus (NPH) versus cerebral atrophy. In patients with the usual clinical triad of dementia, urinary incontinence, and ataxia; computed tomography is a valuable screening test for possible normal pressure hydrocephalus. A CT pattern of enlarged ventricles with normal cortical sulci supports the NPH hypothesis; while a CT pattern of large ventricle plus enlarged cortical sulci supports a diagnosis of cerebral atrophy. The former pattern should be further evaluated with a nuclear medicine CSF flow study; if persistent ventricular penetration of the radiopharmaceutical is demonstrated the NPH diagnosis is likely and the patient has a reasonable chance of improving following a surgical shunting procedure.

CSF ANATOMY AND PHYSIOLOGY The concept of normal CSF circulation is diagramed in Figure 1. The CSF arises primarily from the choroid plexus in lateral, third, and fourth ventricles and flows through and out of the ventricular system into the subarachnoid pathways. CSF flows toward the convexity of the brain where resorption principally takes place through the Pacchionian granulations along the superior sagittal sinus. Part of the CSF descends down the spinal canal then ascends and joins the general circulation. In addition, transmeningeal and transependymal CSF flow has been demonstrated along the spinal canal and ventricles.[9] Circulation of the CSF is summarized in Figure 2.

The CSF space in adults contains 120 to 160 ml of clear fluid and turns over about 5 times a day. The ventricles contain 30 to 40 ml of CSF.

DIAGNOSTIC APPROACH The relative efficacy of CT and radionuclide CSF scanning or both in evaluating dementia is currently under investigation.[1,2] The current diagnostic approach is summarized in Figure 3. Notice that CT is used as a screening test for the demented patient. With the appropriate findings, radionuclide CSF scanning may be the next diagnostic step.

The diagnosis of NPH versus cerebral atrophy is an important one to make because the former may improve with surgical shunting.[3,4]

CSF Imaging Using Computed Tomography

The typical CT pattern in cerebral atrophy is demonstrated in Figure 4. Notice that there is enlargement of both the lateral ventricles and the cortical sulci. In comparison a typical CT pattern in a patient who was diagnosed as normal pressure hydrocephalus and who improved after surgical shunting is shown in Figure 5. Notice here that the ventricles are enlarged and the cortical sulci are normal in size.

CSF Imaging by Nuclear Medicine

Cerebrospinal fluid (CSF) scanning or cisternography is a diagnostic test which includes the injection of an appropriate radiopharmaceutical into the subarachnoid space and sequential images of the head in multiple projections and occasionally of the spinal subarachnoid spaces. This diagnostic procedure is based upon the assumption that the movement and distribution of an appropriate intrathecal radiopharmaceutical reflects bulk CSF flow.[5–7]

Most investigators agree that after the intrathecal injection of an appropriate radiopharmaceutical—includ-

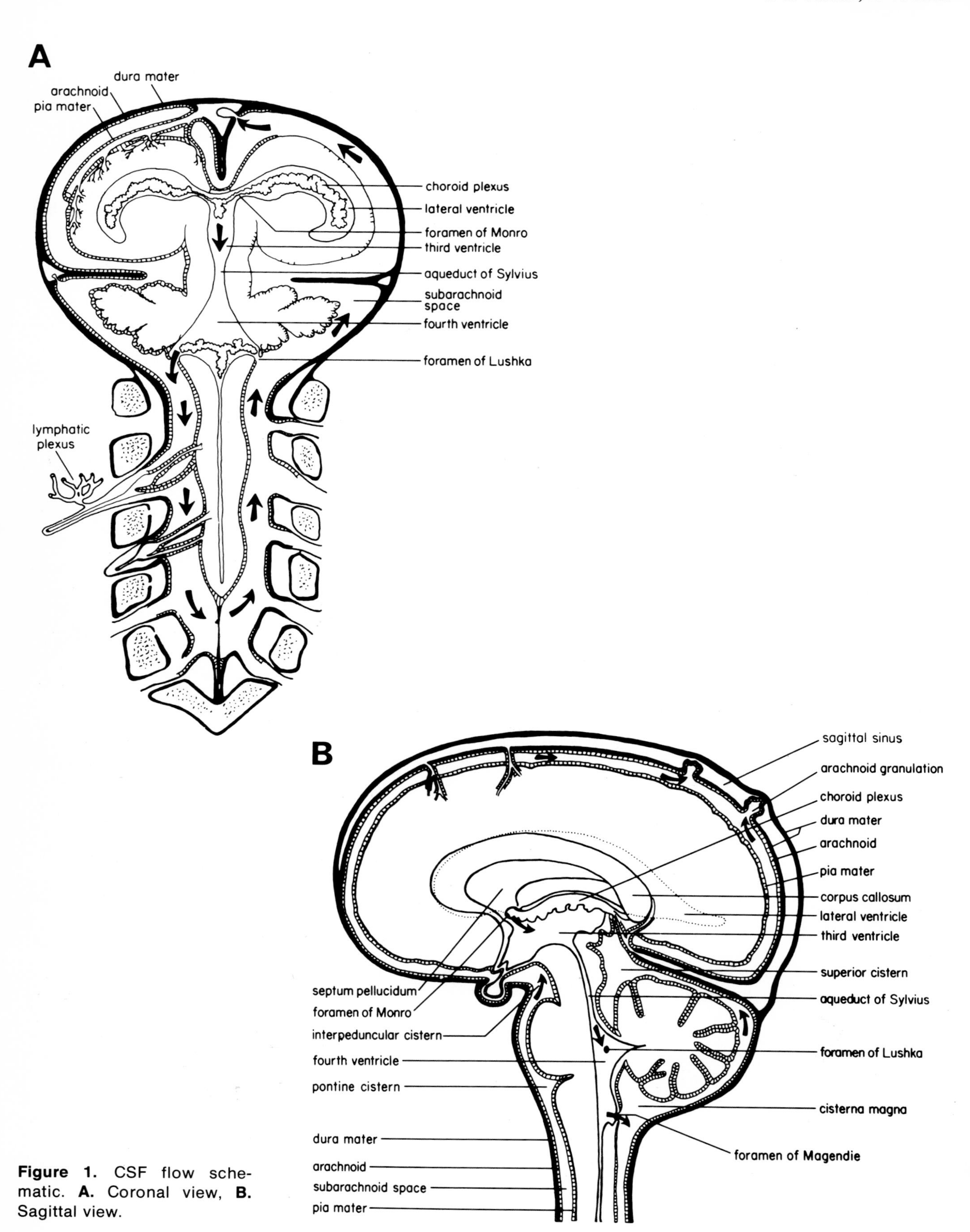

Figure 1. CSF flow schematic. **A.** Coronal view, **B.** Sagittal view.

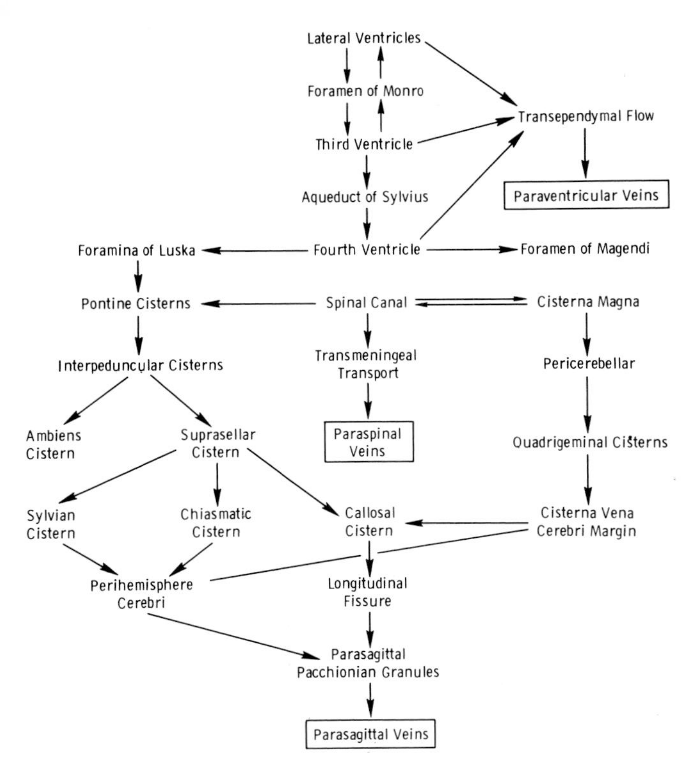

Figure 2. Schematic design of CSF circulation.

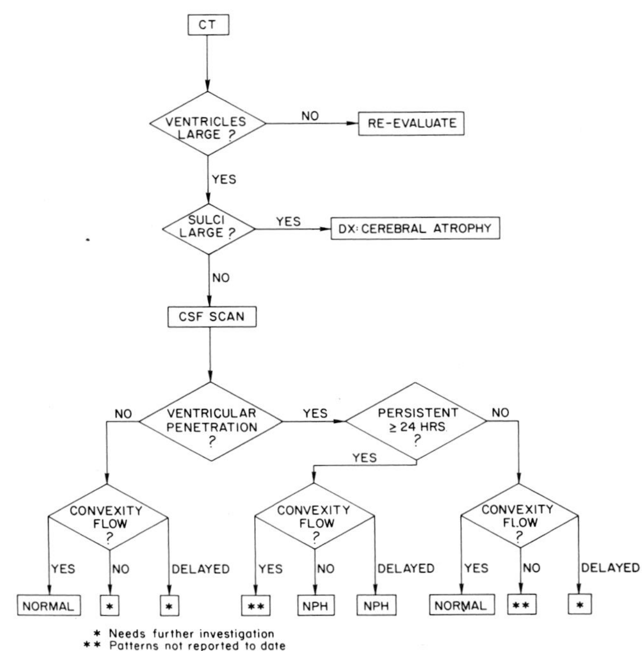

Figure 3. Decision tree for interpretation of CSF images from demented patients.

ing, for example, indium-111 diethylene triaminepentacetic acid (DTPA), ytterbium-169 DTPA, iodine-131 human serum albumin (HSA), gallium-67 DTPA, technetium-99m DTPA, or technetium-99m HSA—the time course of movement accurately depicts CSF flow.[5,8–11] Recently this assumption of correspondence between actual CSF flow patterns with radiopharmaceutical movement in the CSF has been challenged with kinetics studies in dogs and patients.[12] In addition, it has been recently observed, using water-soluble con-

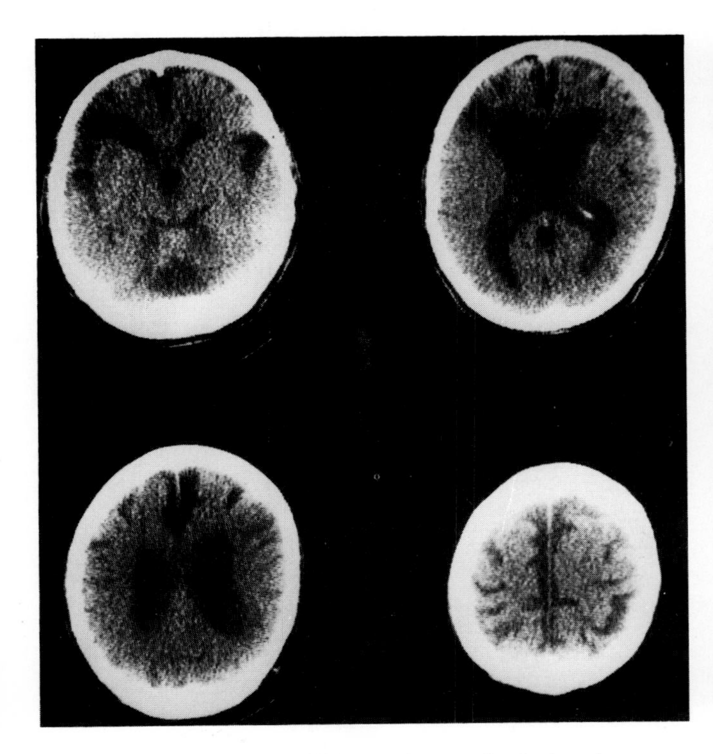

Figure 4. Typical CT scan in cerebral atrophy.

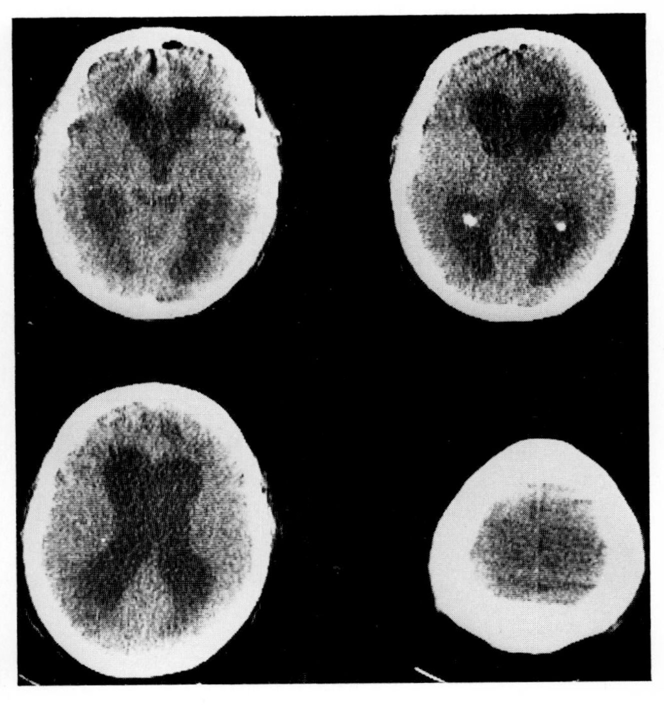

Figure 5. Typical CT scan in normal pressure hydrocephalus.

trast-enhanced CSF studies with computed tomography, that routes of exit of contrast material are dependent upon the molecular weight and size of the contrast molecules; hence, CSF flow studies using contrast-enhanced computed tomography or radiopharmaceuticals are at best an indirect measure of active CSF flow.[13] However, it is generally felt that CSF scanning can give valuable information about treatable forms of hydrocephalus, especially if detailed quantitative flow data in multiple regions are available.

Imaging of the spine following successful introduction of an appropriate radiopharmaceutical into the lumbar subarachnoid space will normally demonstrate a narrow column of activity. By 2 to 4 hours significant activity will ascend to the basal cisterns (more rapid if hyperbaric solution used, where the dose is diluted with an equal volume of 10 percent dextrose in water from a flame-sealed vial). Activity subsequently is apparent in the sylvian cisterns, interhemispheric cistern, and cisterns around the brain stem. Ventricular reflux may take place but is not always appreciated. If this becomes imperceptible or remarkably diminished by 24 hours after injection, it is referred to as transient ventricular filling and considered a normal event.

Activity diminishes with time and appears to migrate over the cerebral cortex to the parasagittal regions with simultaneous loss of activity from the basal cisterns and sylvian regions.

INDICATIONS CSF Scanning may provide useful diagnostic information whenever a disturbance to CSF flow patterns or anatomy is suspected: See, for example, the list in Table 1.

Table 1. Indications for CSF scanning

Detect and characterize hydrocephalus
- Obstructive
- Communicating
- Brain atrophy

Evaluate CSF diversionary shunts
- Ventriculo-peritoneal
- Ventriculo-atrial
- Ventriculo-callosal

Detect and quantify CSF fistulae and extravasation
- Rhinorrhea
- Otorrhea

Characterize localized enlargement of CSF space
- Porencephalic cyst
- Dandy Walker

Detect peripheral, intracranial lesions (rare indications)
- Arnold Chiari
- Subdural hematoma
- Neoplasms
- Cerebral infarcts
- Cerebral ischemic attacks

Research technique to study basic CSF physiology
- Production
- Movement
- Absorption

RADIOPHARMACEUTICALS Commonly used radiopharmaceuticals for CSF scanning are listed in Table 2.

METHODS Following thorough explanation of the procedure to the patient and/or his representative and receiving written consent, a lumbar puncture under aseptic conditions with a number 22 gauge needle is performed.

Scans of the spine may then be made to insure a subarachnoid injection. Multiple views are obtained at 4 to 6 hours, and 24 hours. They include AP, PA, both laterals, and occasionally vertex views. If normal flow is not present, another series of scans at 48 hours may be necessary.

At least 50 K counts should be obtained with each image.[9]

EXAMPLES A normal CSF scan is illustrated in Figure 6. A typical CSF scan in normal pressure hydrocephalus is shown in Figure 7; notice the marked ventricular penetration by the radiopharmaceutical. Varying CSF flow patterns are demonstrated in Figure

Table 2. Radiopharmaceuticals for CSF Scanning[13,14]

Radionuclide	Form	Gamma energy (Kev)	t-1/2	Injection dose μCi*	Radiation dose (rads)	
					Whole body	Brain & cord
^{111}In	DTPA	172,247	2.8 days	500	0.05†	1.2†
^{169}Yb	DTPA	177,198	32 days	500	0.069	8.0‡
^{99m}Tc	HSA or DTPA	140	6 hours	2000	0.032	5.0
^{131}I§	HSA	364	8 days	100	0.17	7.1

* In children a dose of 2–3 μCi/Kg with a minimum of 25 μCi is recommended.
† From Medi-Physics package insert.
‡ Dose to spinal cord and brain surface.
§ No longer used because of high incidence of aseptic meningitis.

logical circumstances following a single injection of tracer so that the effects of changes in afterload, preload, and rate of contraction can be readily evaluated.

In addition, independent measurements of right and left ventricular function are possible.

To date, gated cardiac imaging has proven of value in determining the ejection fraction and end-diastolic volumes of patients with acute myocardial infarction,[7] differentiating aneursym from diffuse hypokinesis of the left ventricle,[8] determining whether right ventricular dysfunction is a cause of low cardiac output in patients with inferior wall infarction,[9] detecting the presence of intraatrial masses,[10] and detecting IHSS,[11] as well as in the pre- and postoperative assessment of ventricular function in patients undergoing coronary bypass surgery.[12]

The ultimate role of multiple gated imaging techniques in the routine evaluation of patients with suspected heart disease is uncertain at present. It is clear that the technique offers an objective measurement of function in each of the cardiac chambers and then can offer information which has previously been obtainable only by cardiac catheterization. The information is significantly different than that available from ultrasound. However, in patients with disorders that cause global changes in ventricular function, it is unlikely that the nuclear technique offers a significant advance over that obtainable with multiprobe or multiphasic echocardiography.

However, in patients with focal disorders of ventricular function, it is likely that nuclear imaging techniques will have significant advantages since the entire heart is visualized in all patients studied. It is likely that the technique will be useful in following patients long-term. This is suggested in the recent evidence of Borer et al. in patients evaluated with aortic regurgitation.[13] Borer and his colleagues found that patients with aortic regurgitation who appeared well compensated at rest could be subdividied into two groups when the ventricular function was measured at rest: (1) those with improvement in ejection fraction; and (2) those with a constant or diminished ejection fraction. In the group with diminished ejection fraction, decompensation occurred clinically over a short period of time; whereas, in those with improved ejection fraction, the clinical course was stable. These data suggest that the time of surgical intervention in patients with aortic regurgitation could be defined by serial measurements of ventricular function at stress.

In patients with congenital disease such as atrial septal defect, Liberthson, et al.[14] have found that serial measurements of right ventricular function may be of value in predicting the optimum time of surgery. Patients with maintained right ventricular function appeared to do well, whereas those with increased right ventricular end-diastolic volume and decreased ejection fraction appeared to have significant morbidity at the time of surgery.

Although the imaging techniques have only been in use for a short period of time, in our clinic they are presently being used to evaluate patients with increasing dyspnea to determine whether this symptom is cardiac in origin. In addition, in patients with known infarction, the gated scan is utilized to separate patients with global ventricular dysfunction from those with more focal abnormalities. The patients with global ventricular dysfunction are considered unsuitable for surgery, whereas those with more focal abnormalities are possible surgical candidates and are usually scheduled for cardiac catheterization as the next step in their preoperative evaluation.

Much of the information from these multigated images is obtained by visually reviewing the computer display of the average cardiac cycle. By simple inspection, regions of abnormal wall motion are readily appreciated and information about global chamber function for each of the chambers can be readily surmised. Thereafter, the semiautomatic approaches to data quantification solidify the subjective impression. Although the technique is relatively new, it appears to warrant a lasting place in the evaluation of ventricular function.

Myocardial Imaging

Global measurements of myocardial blood flow were first made by Love et al. in 1957.[15] This approach was superseded by the selective injection of xenon into the coronary arteries with measurement over the myocardium.[16] Neither measure was able to successfully separate most patients with coronary disease from those without. Since it was obvious that patients with coronary disease had significant alterations in regional myocardial blood flow compared to those without, it was suggested that the technique required to detect patients with regional disease would have to image the myocardium with high spatial resolution. To achieve this goal, Carr and his colleagues first employed the rectilinear scanner and the intravenous administration of cesium-131 to image the regional distribution of myocardial blood flow and to detect areas of myocardial infarction.[17] The diagnosis of myocardial ischemia and the early detection of coronary artery disease, however, was not achieved with this technique until the concept of injecting the radiopharmaceutical at peak stress was advocated.[18,19] This method of injection permitted the evaluation of regional myocardial perfusion reserve and revealed the distribution of regional myocardial perfusion at peak stress. A comparison of this distribution to that obtained following injection with the patient in the basal state revealed areas of the myocardium which became ischemic at stress. Initial comparisons of the stress-injected myocardial perfusion imaging method to that of exercise electrocardiography show an enhanced sensitivity of the imaging technique for the detection of myocardial ischemia. Advances in technology now permit the use of thallium-201 and the scintillation camera for the evaluation of regional myocardial perfusion.[20] The observation of Pohost et al.[21] on the significance of serial imaging following injection at stress permits the separation of patients with myocardial ischemia from those with myocardial scar following a single injection of tracer. Thus, the technique is now greatly simplified from the patient's point of view and an indication of the presence and extent of myocardial ischemia can be achieved in a matter of hours (Fig. 2).

Although it is clear that stress injected myocardial perfusion imaging is more sensitive than that of exercise electrocardiography, it is by no means clear that all patients will benefit from this procedure. The patients who appear to derive greatest benefit at this point are: those with nondiagnostic or uninterpretable electrocardiograms—presence of bundle branch block, treatment with digitalis or patients who develop severe arrhythmias at stress; those patients with markedly abnormal ST segments that are considered due to false positive electrocardiographic responses; and finally those with normal electrocardiographic responses but in whom the clinical suspicion of coronary artery disease is extremely high. In these patients, it is highly likely that the additional cost of stress myocardial imaging would yield a significant benefit. This is particularly true since the tracer technique has a high degree of specificity. An additional group of patients who benefit from the stress injected myocardial perfusion scan are those in whom coronary disease is known to exist, but where a question remains of the sites and extent of myocardial ischemia. This latter application may have only limited benefit since it is likely that patients with multivessel involvement will commonly have one area that becomes ischemic and causes the patient to stop exercising. This area will be the only one defined on the myocardial perfusion scan. Whereas, had the patient continued exercising, it is possible that additional ischemic zones would develop.

In addition to the applications of this method in patients with coronary disease, it has other potential

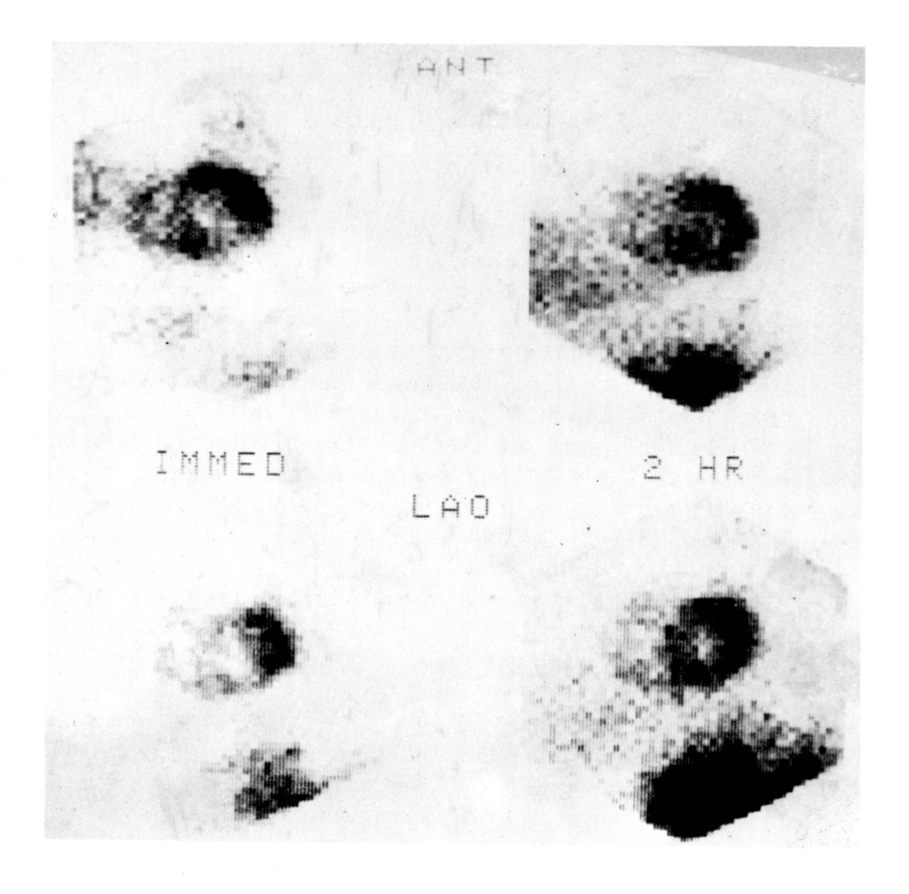

Figure 2. A stress myocardial perfusion scan performed by injection of 1.5 mCi of Thallium-201 chloride at peak treadmill exercise in a 47-year-old man with an 8-month history of increasing angina pectoris. On the initial views taken immediately after stress, there is decreased activity in the inferior wall on the anterior view and in the apex and septum on the LAO view. On the delayed views obtained, 130 minutes later, activity in the inferior and septal and apical walls has increased indicating ischemia. There was significant stenosis demonstrated by angiography of the right coronary and the left anterior descending coronary artery.

applications in cardiomyopathies, disorders involving the right ventricle such as pulmonary hypertension, and possibly in congenital heart disease.

The combination of myocardial perfusion imaging and noninvasive ventricular function measurements make it possible to assess the adequacy of blood flow to a zone of myocardium and the impact of this blood flow on ventricular function. The combination of these two powerful tests should enable more accurate diagnosis of coronary artery disease and its impact on ventricular function.

References

1. Prinzmetal M, Corday E, Spritzler RJ et al.: Radiocardiography and Its Clinical Applications, JAMA 139:617, 1949
2. Folse R, Braunwald E: Determination of Fraction of End Diastolic Volume Ejected per Beat and of Ventricular End Diastolic and Residual Volumes, Circulation 25:674, 1962
3. Mullins CB, Mason DT, Ashburn WL et al.: Determination of Ventricular Volume by Radioisotope Angiography, Am J Cardiol 24:72, 1969
4. Strauss HW, Zaret BL, Hurley PJ: A Scintiphotographic Method for Measuring Left Ventricular Ejection Fraction in Man without Cardiac Catheterization, Am J Cardiol 28:575, 1971
5. Borer JS, Bacharach SL, Green MV et al.: Real-Time Radionuclide Cineangiography in the Noninvasive Evaluation of Global and Regional Left Ventricular Function at Rest and During Exercise in Patients with Coronary-Artery Disease, New Engl J Med 296:839, 1977
6. Burow R, Strauss HW, Singleton R, et al.: Analysis of Left Ventricular Function from Multiple Gated Acquisition (MUGA) Cardiac Blood Pool Imaging: Comparison to Contrast Angiography, Circulation 56:1024, 1977
7. Zaret BL, Strauss HW, Hurley PJ et al.: A Noninvasive Scintiphotographic Method for Detection of Regional Ventricular Dysfunction in Man, New Engl J Med 284:1165, 1971.
8. Rigo P, Strauss GW, Taylor D, et al.: Left Ventricular Function in Acute Myocardial Infarction Evaluated by Gated Scintiphotography, Circulation 60:67, 1974
9. Rigo P, Murray M, Taylor DR, et al.: Right Ventricular Dysfunction Detected by Gated Scintiphotography in Patients with Acute Inferior Infarction, Circulation 52;268, 1975
10. Pohost GM, Pastore JO, McKusick KA, et al.: Detecting Left Atrial Myxoma by Gated Radionuclide Imaging, Circulation 55:88, 1977
11. Pohost GM, Vignola PA, McKusick KA, et al.: Hypertrophic Cardiomyopathy: Evaluation by Gated Cardiac Blood Pool Scanning, Circulation 55:92, 1977
12. Cohen H: Personal Communication
13. Borer JS, Bacharach SL, Green MV: Left Ventricular Function During Exercise Before and After Aortic Valve Replacement, Circulation 55&56 (Suppl III): III–28, 1977
14. Liberthson R: Personal Communication
15. Love WD, Burch GE: A Study in Dogs of Methods Notable for Estimating the Rate of Myocardial Uptake of ^{86}Rb in Man and the Effect of L-Norepinephrine and Pirtessin on ^{86}Rb Uptake, J Clin Invest 36:468, 1957
16. Ross RS: Myocardial Perfusion—Historical Perspectives and Future Needs. In Strauss HW, Pitt B, James A (eds): Cardiovascular Nuclear Medicine. St Louis, Mosby, 1974
17. Carr EA, Gleason F, Shaw J, et al.: The Direct Diagnosis of Myocardial Infarction by Photoscanning After Administration of Cesium-131, Am Heart J 68:627, 1964
18. Strauss HW, Zaret BL, Martin ND, et al.: Noninvasive Evaluation of Regional Myocardial Perfusion with Potassium-43: Technique in Patients with Exercise Induced Transient Myocardial Ischemia, Radiology 108:85, 1973
19. Zaret BL, Strauss HW, Martin ND, et al.: Noninvasive Regional Myocardial Perfusion with Radioactive Potassium: Study of Patients at Rest, with Excercise and during Angina Pectoris, New Engl J Med 288:809, 1973
20. Bailey IK, Griffith LSC, Rouleau J, et al.: Thallium-201 Myocardial Perfusion Imaging at Rest and During Excercise: Comparative Sensitivity to Electrocardiography in Coronary Artery Disease, Circulation 55:79, 1977
21. Pohost GM, Zir LM, Moore RH, et al.: Differentiation of Transiently Ischemic from Infarcted Myocardium by Serial Imaging After a Single Dose of Thallium-201, Circulation 55:294, 1977

Discussion: Cardiac Imaging

Moderator: H. William Strauss
Panelist: Walter L. Henry

STRAUSS: Please describe your methods of computing the area of the mitral orifice and how you use the one-centimeter marks.

HENRY: We take a video tape which is being displayed on a video monitor and gradually step through frame by frame until we choose a frame in early diastole in which the outline of the mitral orifice is clearly seen. At that point we place a piece of transparent paper over the top of the image, take a wax pen and actually draw on the image. We draw a line just peripheral to the inner boundary outlining the mitral orifice. You can take a small device and chronometer that area, or you can simply lay it over the top of the piece of graph paper that is marked off in small boxes as I do. I simply count the number of small boxes within that outline as it is traced. And I do the same thing for a one-centimeter calibration marker. I draw a one-centimeter calibration marker and I figure out how many small boxes would fit inside one square centimeter area and then use the two to compute the mitral orifice area.

COMMENT: Computing the mitral orifice area is not as easy as it sounds because you really have to be sure that you are down at the smallest part of the mitral valve orifice. If you make that measurement up high in the part of the anterior leaflet that is bulging, you are going to get an artifically high measurement. So you must be very careful about angling your transducers down to the smallest part of the mitral orifice just above the papillary muscles.

HENRY: There are some papers out that describe those techniques in some detail. Basically we first image the papillary muscle and then gradually scan from the papillary muscle up toward the mitral valve. The first orifice we see is what we chronometer.

QUESTION: Can thallium studies be done without computers, or really what are the equipment requirements for cardiac work?

STRAUSS: I think that one could certainly get into it with a scintillation camera recording data on film. Dr. Henry and Dr. Winsberg have indicated that you are really better off with an echo machine which can do something in addition to M mode, such as an arc echo or a phased array when doing echocardiography. I think the same is true with the nuclear techniques in that you are far better off with a computer to alter contrast and brightness. The interpretation of the images are made with much greater certainty if you have a computer system. Certainly Dr. Mishkin, for example, has succeeded very well in recording gated scans without the use of a computer system, as we did when we first started. However, the improvement in the certainty of diagnosis is so great with a computer that I do not recommend that anybody try to do it the way we did in 1969.

I think the same is true of thallium imaging. It is clear that you can very easily record a thallium scan without a computer on a piece of film. Unfortunately, these images are very difficult to interpret.

I think the studies require a lot of physician participation, both in their performance and in their interpretation.

With reference to the thallium imaging it is mandatory, I believe, to have a cardiologist actually supervise the stress test. In our institution the stress test is supervised by a full-time cardiologist who is present in the lab and does all the stress tests for us. This is also true with the stress gated studies.

QUESTION: Why does an MI present as a cold defect on thallium and a hot defect on pyrophosphate?

STRAUSS: The thallium is distributed in the myocardium in relation to blood flow. So in an area where there is no blood flow there is no thallium, and hence the cold spot. With pyrophosphate imaging, on the other hand, one needs a damaged cell membrane to permit the material to enter the cell. Normal cells exclude it. So when you have damaged tissue then and only then can the pyrophosphate enter. There has been a lot of controversy about how much damage you need, whether the cell must be irreversibly damaged or not.

I think all the evidence at present indicates that the cell must be at least on death's doorstep if not absolutely dead for the pyrophosphate to enter. Hence we have this difference with pyrophosphate. We will see a hot spot because it is a severely damaged cell membrane or because the cell is dying or dead, and with thallium we will see a cold spot for the same reason.

STRAUSS: I think the data that have been obtained thus far suggest that the exercise gating soon will be more sensitive certainly than the stress electrocardiogram and I think probably more sensitive than the thallium scan. I think there are a number of groups now investigating that specific question.

The sensitivity of any test is directly related to the population studied. If you take a population of patients, all of whom complain of the classic chest pain..."I get the pain in my chest and it goes down my arm and it comes on when I exercise," and if you do stress electrocardiograms on that subset of patients, the tests will be exquisitely sensitive and will show a 95 percent correct diagnosis of coronary disease. On the other hand, if you do what Jeff Borer did a number of years ago, and just test people at random to find out if they have coronary disease, then you find out that the specificity of the stress electrocardiogram test is really very poor.

As I recall, only about 45 to 50 percent were correctly diagnosed, and the number of false positives was about 40 percent. I think it is who you study that will make the test look better or worse. The same is absolutely true of the thallium: if you take patients with fairly clearcut symptomatic coronary disease, the thallium is going to look spectacular in detecting it. On the other hand, when you go to the local health salon and do thallium scans on all people before they enter your exercise program, it is going to look terrible.

Value of Tomographic Imaging of Gallium Citrate for Tumor and Inflammatory Process Localization*

Wayne W. Wenzel

Emission tomography has been a topic of interest in the literature as early as 1963[1] when Kuhl and Edwards described their method of image separation in radioisotope scanning. In 1966, Hal Anger first published his idea of simultaneous readout planar tomography which has become the first clinically used tool to perform emission tomography. His basic design (Fig. 1) is built upon the Anger camera concept with multiple photomultiplier tubes, a large field of view crystal, and XY logics for scintillation event localization. To this (Fig. 2), he has added the desirable functions of the rectilinear scanner, focused collimation, slightly thicker crystal for increased efficiency, and additional XY logics inherent to the rectilinear motion of scanning. The tomographic scanner (Fig. 3) then, is a culmination of a multiple photomultiplier tube, large field of view crystal, focused collimator, and rectilinear scanning system in motion over the patient. The finished product is composed of upper and lower probes to increase efficiency.

Principle of Operation

Perhaps the best way of explaining the way this tomographic imager works is to study the field of view of a focused collimator (Fig. 4). One will note that Figure 5 shows in cross-section the approximated planes described in Figure 4. The field of view of the focused collimator is largest near its surface and decreases at a regular rate to its geometric focal point at which point it is theoretically zero. It then enlarges again in a neg-

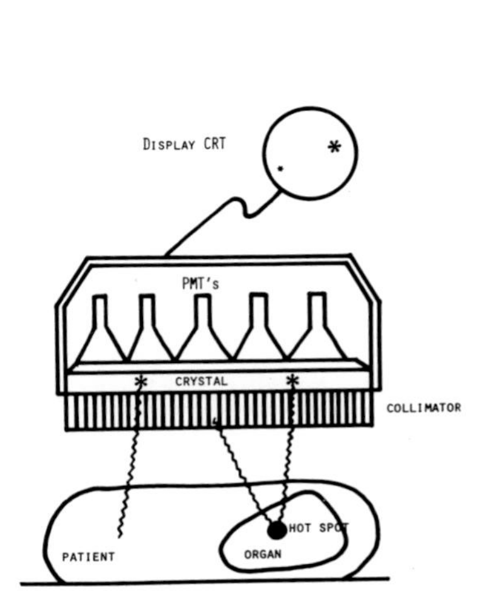

Figure 1. Anger camera.

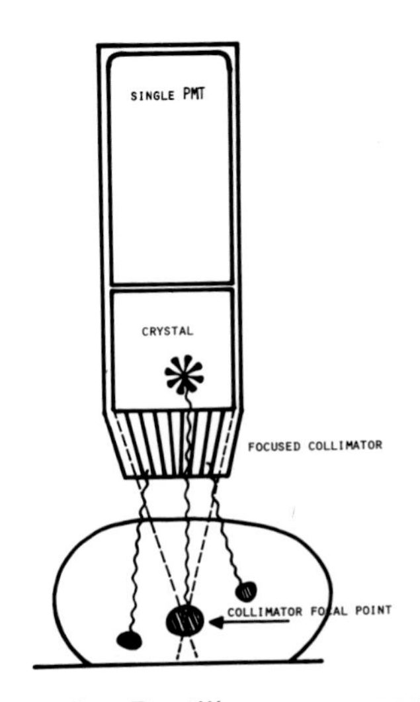

Figure 2. Rectilinear scanner.

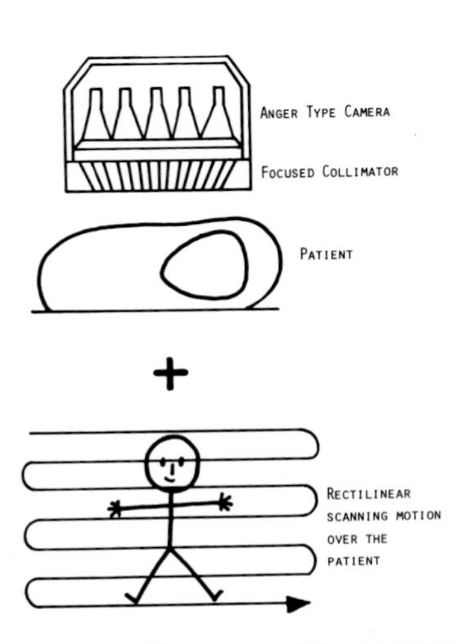

Figure 3. Tomographic scanner.

* This chapter first appeared under the title of "Clinical Applications of the Pho/Con Multiplane Tomographic Scanner: Value of Tomographic Imaging of Gallium Citrate for Tumor and Inflammatory Process Localization" published by Searle Radiographics, a division of Searle Diagnostics, Inc., Des Plaines, Illinois and is reprinted here with permission.

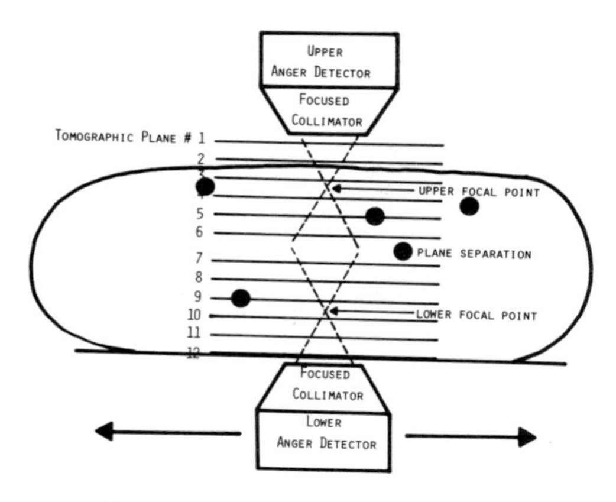

Figure 4. Plane specification.

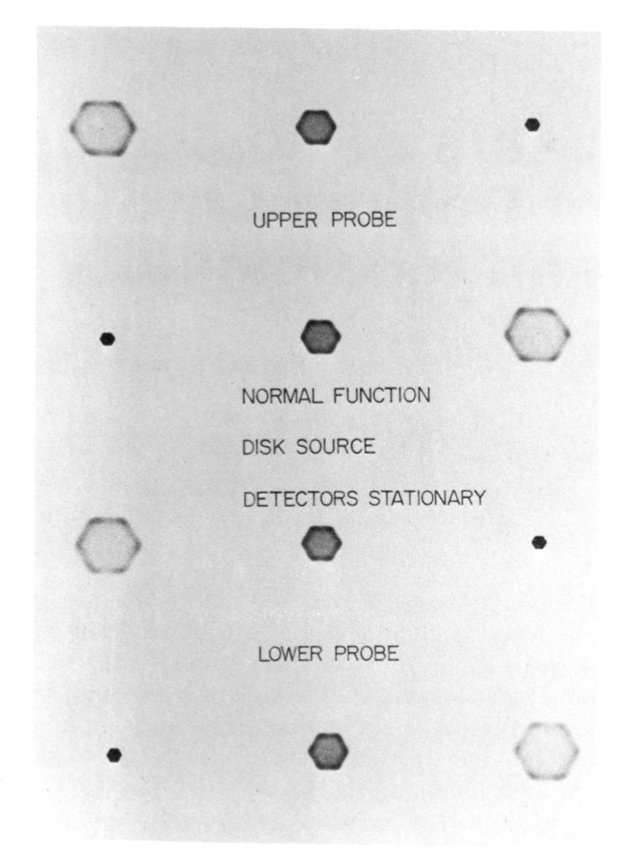

Figure 5.

ative fashion, becoming larger instead of smaller. Several basic principles are true regarding point sources or foci of radioactivity within a patient that come into and out of the field of view of such a collimator. The first basic principle is that sources located above the geometric focal point of the collimator will produce images on the cathode ray display of the Anger camera which move across that cathode ray tube in the opposite direction of the detector motion (Fig. 6). Sources located below the geometric focal point of the collimator will produce images on the cathode ray tube which move across the display in the same direction as the detector motion (Fig. 7).

The third basic principle is that the speed of the image motion across the cathode ray tube, or the time during which the point source is imaged, varies as the source distance from the geometric focal point of the collimator. In other words, the smaller the field of view of the detector, the shorter the visualization time or the faster the source will move in and out of view. This principle applies to sources located both above and below the geometric focal point—see Figure 4. Now, if the photographic film used to record the image is moved across the CRT in the same direction and at the same speed as the movement of the source's image, then that source and all other sources in that plane will be sharply imaged. Images from other planes which are moving at different speeds or directions will be blurred on the film and, thus, not imaged. The necessary movement can be achieved either mechanically or by electronic means. Finally, to image any particular plane, all that is necessary is to synchronize the movement of the film either mechanically or electronically to the movement of that plane's images—see Figure 5. The Pho/Con™ tomographic scanner, as designed by Hal Anger, accomplishes this feat electronically by varying the size of the image and the direction of motion electronically. Twelve planes are simultaneously processed electroni-

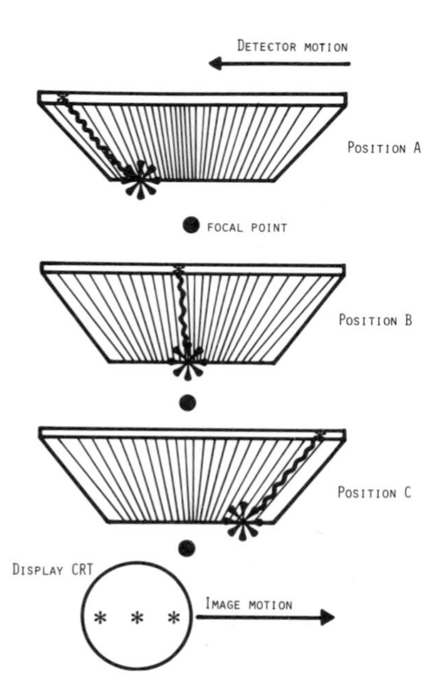

Figure 6. Image formation from sources above focal point.

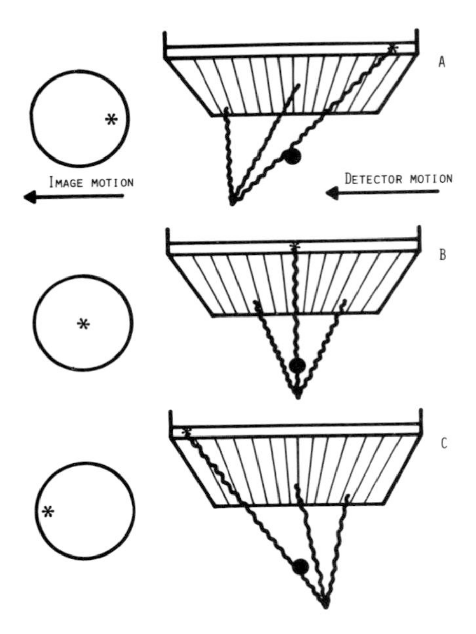

Figure 7. Image formation from sources below focal point.

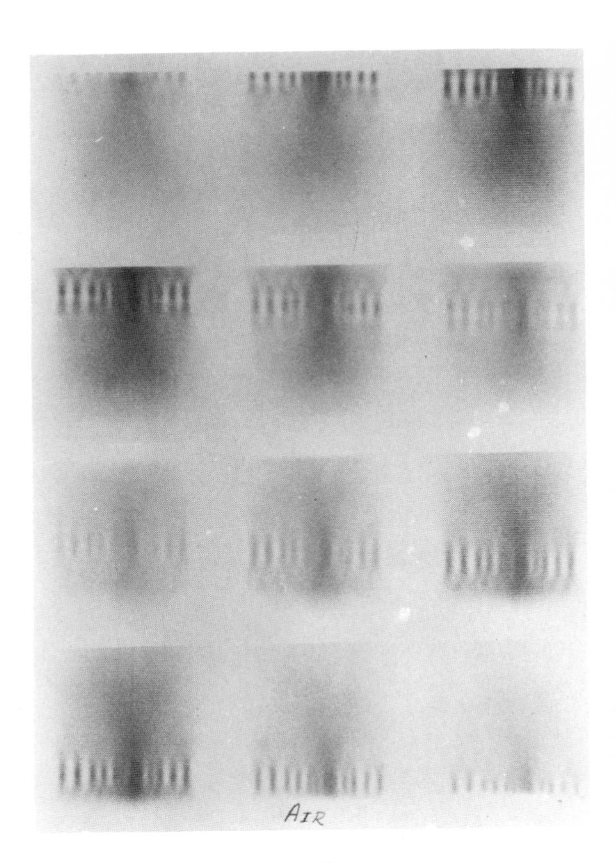

Figure 8.

cally and displayed, though the number of planes is theoretically infinite.

Tomography is accomplished by this technique and is nicely demonstrated by a 45-degree bar phantom designed by Jerry Hines and the Bureau of Radiological Health (Fig. 8).

The tool is theoretically and realistically proven with phantom studies to this point, but most important is whether this tool is clinically useful. Gallium imaging is realistically an area which needs improvement and theoretically could be improved by tomographic imaging as planar information would seem desirable. We have had the opportunity to use this instrument for 14 months in the clinical setting. We have found the tomographic imager to be a superior piece of equipment for gallium imaging, and I will present our reasoning in the form of clinical examples.

Clinical Applications

The first is an 8-year-old youngster with rhabdomyosarcoma. Previous gallium images showed her tumor to be gallium positive and we were performing this examination as a form of follow-up. This anterior view of the chest taken with an Anger-type scintillation camera is felt to be within normal limits (Fig. 9). However, tomographic images reveal a definite mediastinal uptake (Fig. 10) in recurrent tumor, even in the face of negative chest x-ray and negative scintillation camera image (Fig 9). This represents an example where tomography exclusively demonstrated the lesion. In the

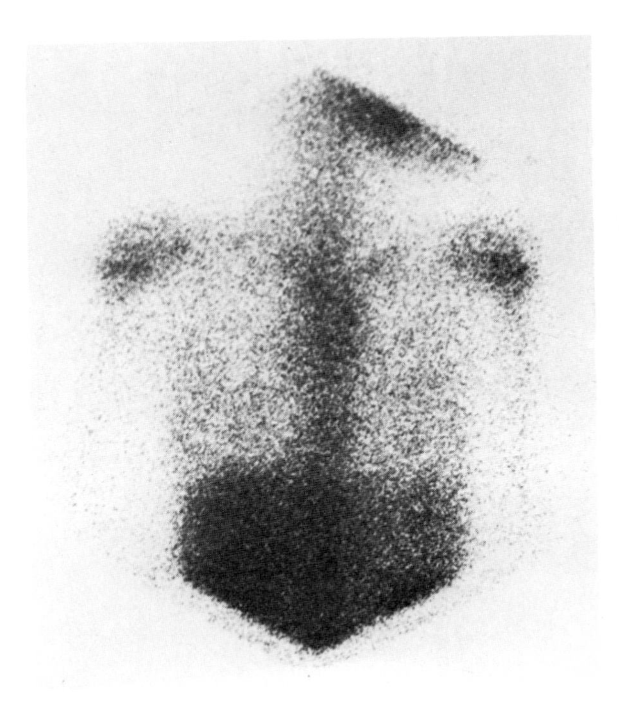

Figure 9.

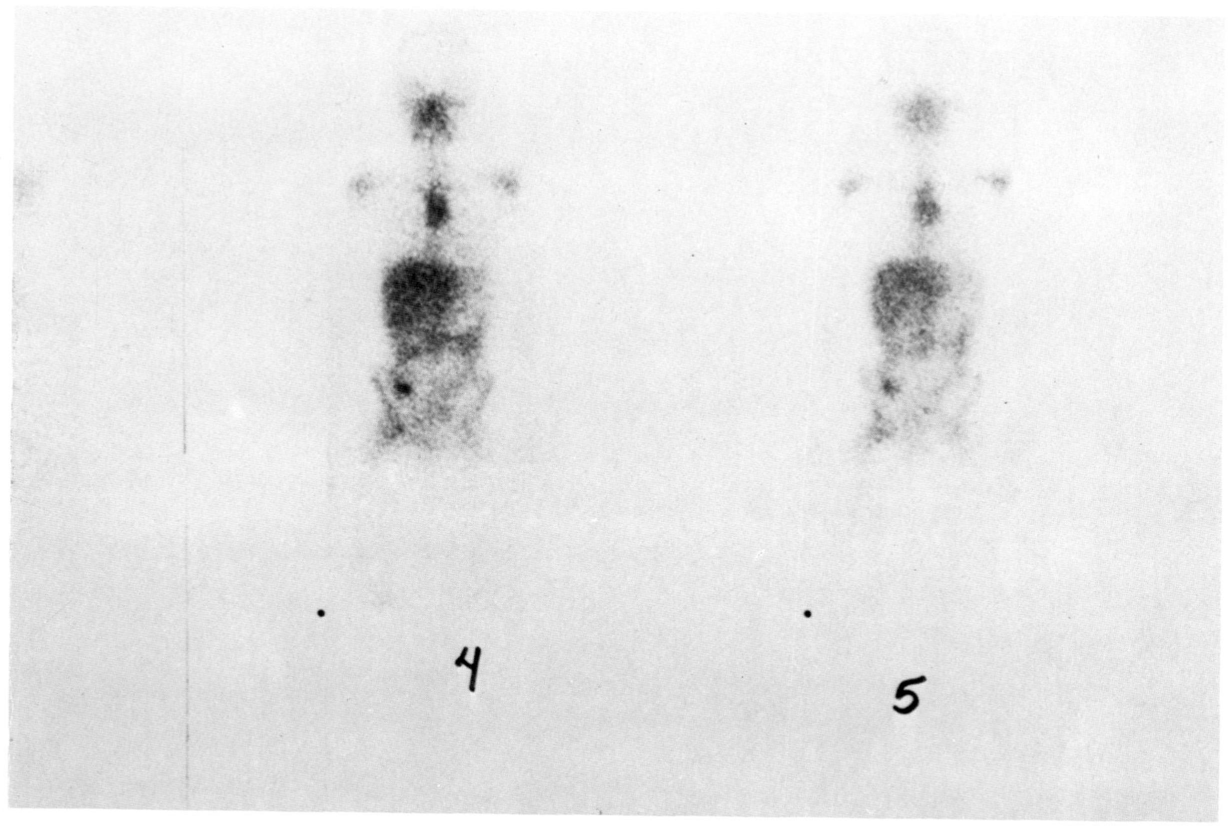

Figure 10.

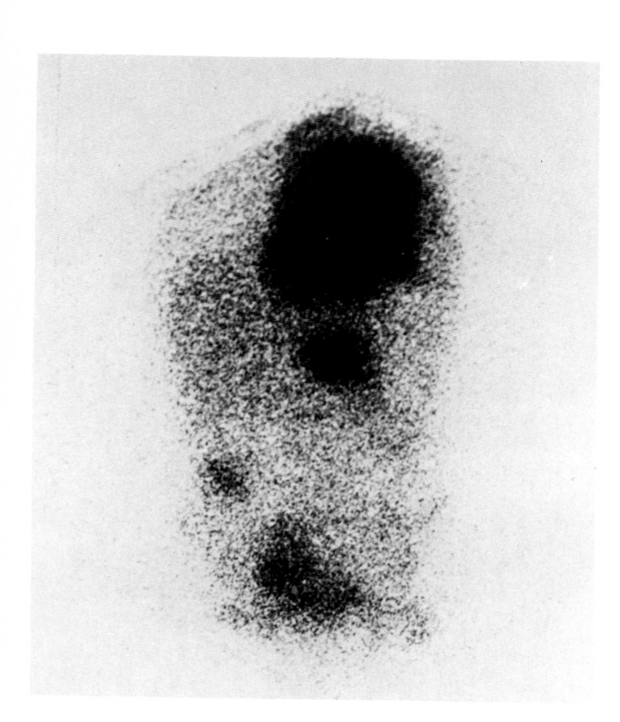

Figure 11.

majority of cases, tomography serves to better delineate lesions, but not necessarily to show otherwise nondetectable lesions.

The next case is an example where the Anger camera anterior view of a 3-year-old child with Burkitt's lymphoma shows multiple lesions below the diaphragm and one large lesion in the left hemithorax (Fig. 11).

Tomographic images nicely separate these lesions into different planes. Gallium scanning has been reported to show lesions not detectable by chest radiographs. Having tomographic capability and presentation further enhances this capability and allows one to more precisely place the level of lesion for subsequent biopsy (Fig. 12). Such a case is this patient with adenocarcinoma of the uterus, recurrent in the pelvis. Her entire work-up, including a liver scan, was negative, but gallium scan revealed the lesion in the pelvis, a lesion in the liver anteriorly, and a right sided mediastinal lesion. Retrospective evaluation of the liver scan did not reveal this lesion, and chest x-ray was negative. Exploratory surgery of the abdomen revealed a thin "plaque-like" focus of tumor on the dome of the liver anteriorly and subsequently the mediastinal lesion has become apparent on chest radiographs. This case illustrates not only the extreme value of gallium imaging in staging of tumors, but also the value of the tomographic images

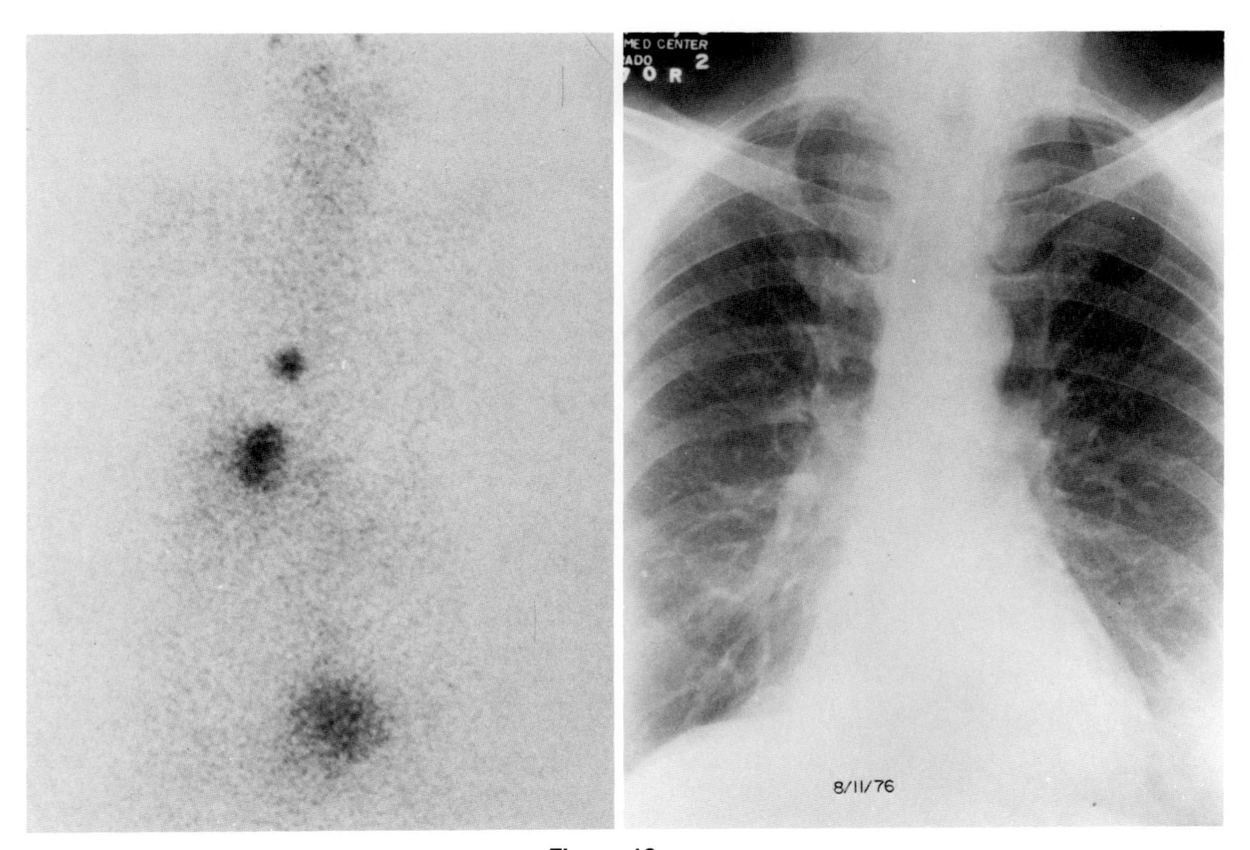

Figure 12.

which directed the surgeon to the proper place for biopsy confirmation.

The next patient was suspected of having a malignancy by the referring physician six months prior to this examination (Fig 13a.). Total "routine" work-up was accomplished and no abnormalities were found. The patient returned after six months with an additional weight loss of 20 pounds for reevaluation. Gallium scan was accomplished at this time after, again, a totally negative "routine" work-up, and revealed abnormal uptake presumably in neoplasm in the mid- and upper epigastrium, superior mediastinum on the left, and left supraclavicular area. Since this was the only evidence of abnormality, it directed surgical exploration to the supraclavicular region where diagnosis of metastatic adenocarcinoma was obtained. Ultrasound evaluation (see Fig. 13b) revealed the uptake in the mid-epigastrium to be adenopathy adjacent to a large

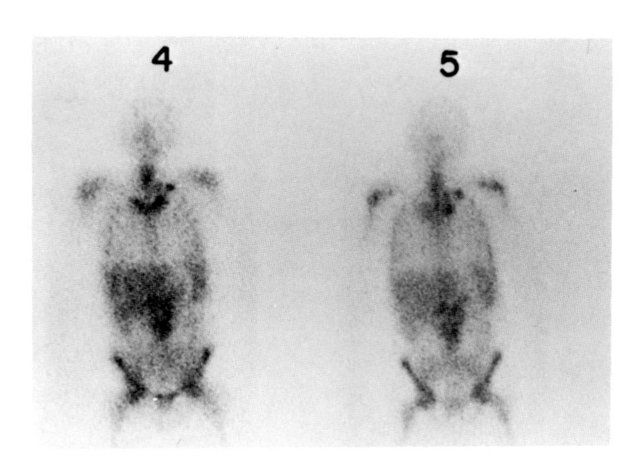

Figure 13a.

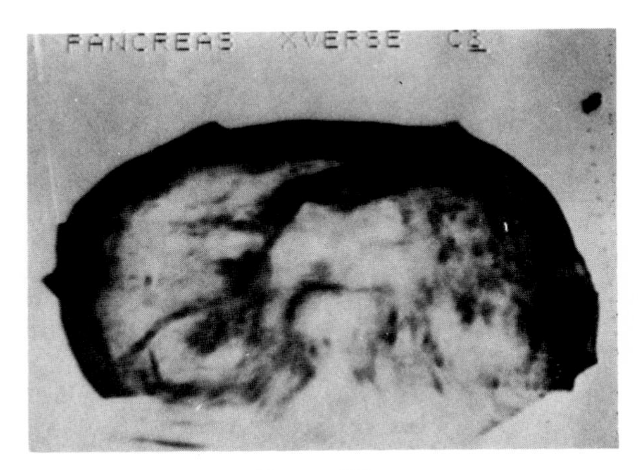

Figure 13b.

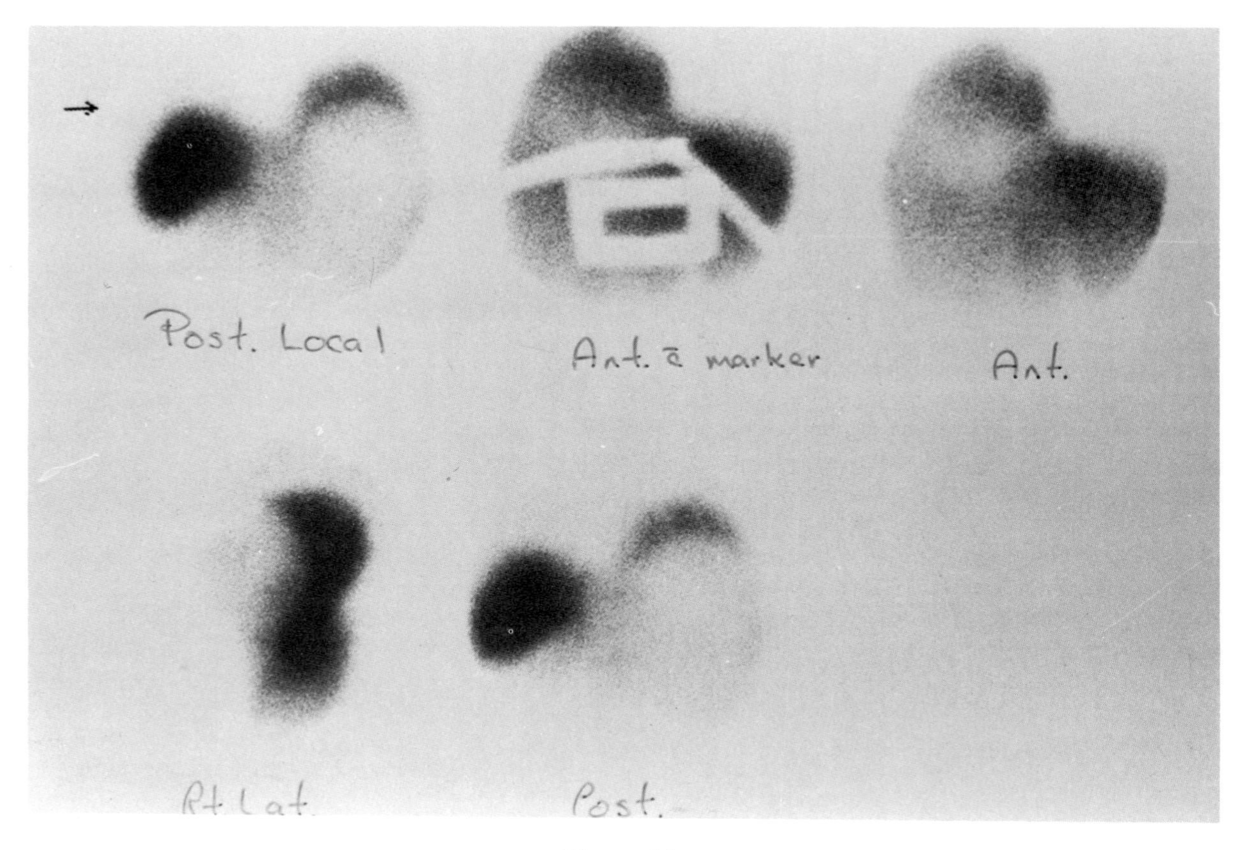

Figure 14.

pancreatic mass which was confirmed at surgery: again, an example of the value of gallium imaging, especially when linked with tomographic depth information.

Gallium is also proven to be invaluable in location of inflammatory processes. Since inflammatory processes are frequently abscesses and need surgical exploration and draining, and since gallium scanning is frequently the only examination which shows the focus of abscess, precise determination of location within the body is of extreme value to the surgeon. The next case shows an obvious lesion on the liver scan showing a large posterior filling defect (Fig. 14). Figure 15 shows tomographic gallium images which reveal a rim of increased uptake around a relatively cool "photogenic" area in the same region as the filling defect on the liver scan, and a wedge-shaped area of increased uptake in pleural

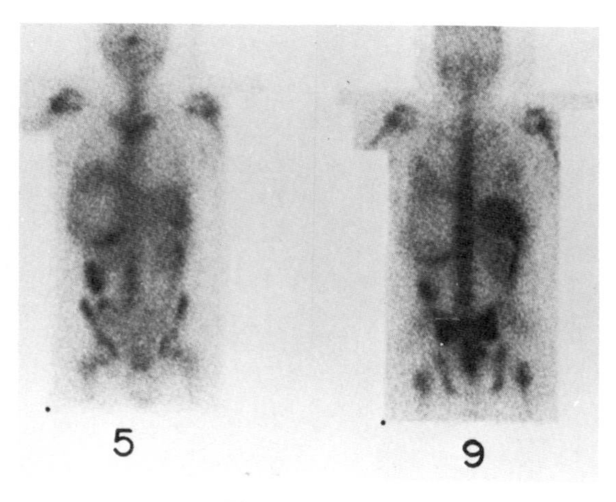

Figure 15.

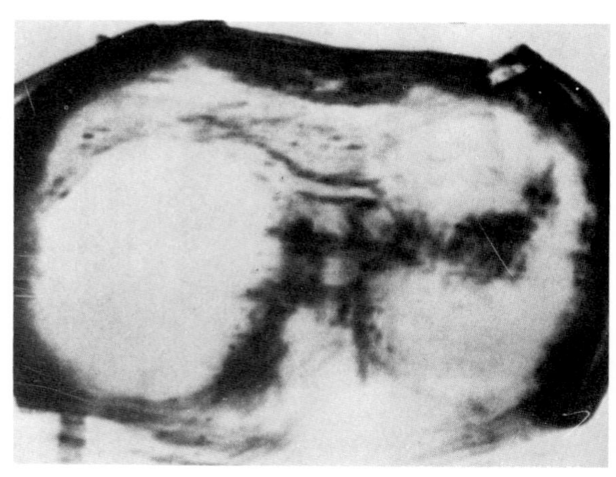

Figure 16.

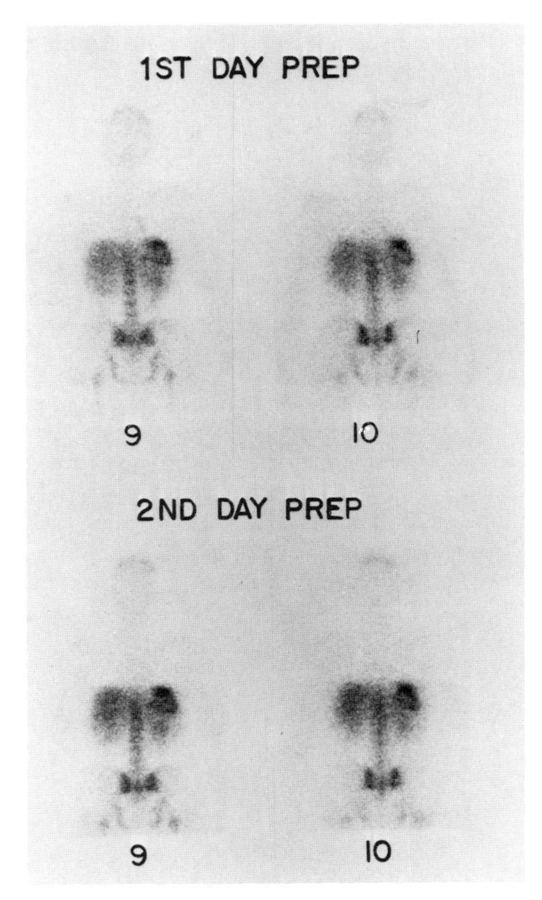

Figure 17.

fluid in the right hemithorax posteriorly. Linking this information with ultrasound information (Fig. 16) shows this lesion to be cystic in nature and cinches the diagnosis of abscess in this patient.

Less obvious, however, is this next case in a young man with Hodgkin's lymphoma who had undergone a staging laparotomy before radiation and chemotherapy. He presented to us with a fever of unknown origin and recurrent lymphoma was highly suspected. Tomographic gallium images revealed plaquelike increased activity (which did not change after repreparation of the bowel) in the splenic fossa conforming to the shape of the diaphragm (Fig. 17). The ability to localize this lesion as being posterior in nature was invaluable in this case as it helped direct the biopsy needle which revealed subphrenic abscess.

Conclusion

In conclusion, the Pho/Con™ tomographic scanner has undergone extensive clinical evaluation and is a proven clinically useful tool to perform emission tomography. By its very nature, gallium scintigraphy lends itself nicely to tomographic imaging and has proven itself invaluable in more precise localization of neoplastic and inflammatory abnormalities.

Acknowledgments

The authors would like to thank Dr. Jerry Hines and Mr. Gordon Kenny for physics support and phantom studies, Miss Senja Erickson for illustrations, and Mrs. Susan Roberdes for preparation of this manuscript.

References

1. Kuhl DE, Edwards RQ: Image Separation Radioisotope Scanning, Radiology 80:653, 1963
2. Kuhl DE, Edwards RQ: Cylindrical and Section Radioisotope Scanning of the Liver and Brain, Radiology 83:926, 1964
3. Bozzo SR, Robertson JS, Milazzo JP: A Data Processing Method for a Multidetector Postion Scanner. In Gottschalk A, Beck RN (eds): Fundamental Problems in Scanning, Springfield, Ill, Thomas, 1968, p. 212
4. Harper PV: The Three-Dimensional Reconstruction of Isotope Distributions. In Gottschalk A, Beck RN (eds): Fundamental Problems in Scanning, Springfield, Ill, Thomas, 1968, p 191
5. Hisada K, Hiraki T, Ohba S, Matsudaira M: Simultaneous Performance of Isosensitive Scanning and Bilaminoscanning, Radiology 88:129, 1967
6. Cassen B: Problems of Quantum Utilization. In Gottschalk A, Beck RN (eds): Fundamental Problems in Scanning, Springfield, Ill, Thomas, 1968, p. 50
7. Crandall, P, Cassen B: High Speed Section Scanning of the Brain, Arch Neurol 15:163, 1966
8. McAfee JG, Mozley JM, Natarajan TK, Fueger GF, Wagner HN Jr: Scintillation Scanning with an Eight-Inch Diameter Sodium Iodide (Tl) Crystal, J Nucl Med 7:521, 1966
9. Hart HE: Focusing Collimator Coincidence Scanning, Radiology 84:126, 1965
10. Mozley JM: Unpublished Data
11. Anger HO: Survey of Radioisotope Cameras, ISA Trans 5:311, 1966
12. Anger HO: Tomographic Gamma-Ray Scanner with Simultaneous Readout of Several Planes, U.C.R.L.-16899, May, 1966
13. Anger HO: Tomographic Gamma-Ray Scanner with Simultaneous Readout of Several Planes. In Gottschalk A, Beck RN (eds): Fundamental Problems in Scanning, Springfield, Ill, Thomas, 1968, p. 195
14. Cassen B: Image Formation by Electronic Crosstime Correlation of Signals from Angular Ranges of Unfocused Collimating Channels. In Medical Radioisotope Scintigraphy, Vol. 1, IAEA, Vienna, 1969, p. 261
15. McAfee JG, Mozley JM, Stabler EP: Longitudinal Tomographic Radioisotopic Imaging with a Scintillation Camera: Theoretical Consideration of a New Method, J Nucl Med 10:654, 1969

16. Andrews GA, Edwards CL: Tumor Scanning with Gallium-67, JAMA 233:1100, 1975
17. Clausen J, Edeling CJ, Fogh J: ^{67}Ga Binding to Human Serum Proteins and Tumor Components, Cancer Res 34:1931, 1974
18. Higasi T, Nakayama Y, Murata A, Nakamura K et al: Clinical Evaluation of ^{67}Ga-citrate Scanning, J Nucl Med 13:196, 1972
19. Johnston GS, Jones AE, Milder MS, Frankel RL: The Gallium-67 Scan in Clinical Assessment of Cancer, J Surg Oncol 5:529,1973
20. Kramer RJ, Larson SM, Milder MS, Herdt JR et al: Localization of Gallium-67 Citrate in Unsuspected Sites of Neoplastic Disease, IAEA SM-164/150:641, 1972
21. Larson SM, Milder MS, Johnston GS: Interpretation of the ^{67}Gallium Photoscan, J Nucl Med 14:208, 1972
22. Littenberg RL, Alazraki NP, Taketa RM, Reit R et al: A Clinical Evaluation of Gallium-67 Citrate Scanning, Surg Gyn Obst 137:424, 1973
23. Henkin RE, Quinn JL, Weinberg PE: Adjunctive Brain Scanning with ^{67}Ga in Metastases, Radiology 106:595, 1973
24. Jones AE, Frankel RS, DiChiro G, Johnston, GS: Brain Scintigraphy with ^{99m}Tc Pertechnetate, ^{99m}Tc Polyphosphate, and ^{67}Ga Citrate, Radiology 112:123, 1974
25. Waxman AD, Siemsen JK, Lee GC, Wolfstein RS et al: Reliability of Gallium Brain Scanning in the Detection and Differentiation of Central Nervous System Lesions, Radiology 116:675, 1975
26. Kashima HK, McKusick KA, Malmud LS, Wagner HN Jr: Gallium-67 Scanning in Patients with Head and Neck Cancer, Laryngoscope 84:1078, 1974
27. Silberstein EB, Kornblut A, Shumrick DA, Saenger EL: ^{67}Ga as a Diagnostic Agent for the Detection of Head and Neck Tumors and Lymphoma, Radiology 110:605, 1974
28. Cellerino A, Filippi PG, Chiantarello A, Borasio P: Operative and Pathologic Survey of 50 Cases of Peripheral Lung Tumors Scanned with 67-Gallium, Chest 64:700, 1973
29. van der Schoot JB, Groen AS, de Jong J: Gallium-67 Scintigraphy in Lung Diseases, Thorax 27:543, 1972
30. Henkin RE: Gallium-67 Citrate Scanning in Malignant Lymphoma, CRC Crit Rev Clin Radiol Nucl Med 7:65, 1975
31. Richman SD, Levenson SM, Jones AE, Johnston GS: Radionuclide Studies in Hodgkin's Disease and Lymphomas, Sem Nucl Med 5:103, 1975
32. Milder MS, Frankel RS, Bulkley GB, Ketcham AS, et al: Gallium-67 Scintigraphy in Malignant Melanoma, Cancer 32:1350, 1973
33. Bailey TB, Pinsky SM, Mittenmeyer BT, Borski AA, et al: A New Adjuvant in Testis Tumor Staging: Gallium-67 Citrate, J Urol 110:307, 1973

Discussion: Gallium-67 and Bone Scanning

Moderator: Thomas A. Verdon
Panelists: Wayne W. Wenzel
Ernest Fordham

VERDON: Dr. Wenzel, you mentioned the problem of oat cell carcinoma. Do they really concentrate gallium-67?

WENZEL: Yes, they do, but we don't have the opportunity to scan many of the oat cell carcinomas with gallium at our institution. Scanning bronchogenic carcinomas with gallium is really pretty successful. Excellent results have been published in the literature by several people and our experience is really quite good, but not 100 percent accurate, as has been reported.

FORDHAM: Yes, oat cell uptake is not as good as that of the epidermoid bronchogenic cancer, so our yield in the oat cell carcinoma is not going to be as high as the tremendously high uptake in epidermoid bronchogenic carcinoma. The areas that I think gallium scanning is useful in are, of course, the lymphomas, particularly Hodgkins, and histiocytic lymphoma, where we have very high pickup rates. Melanoma pickup is extremely high and at our institution, it is considered malpractice to do anything with a melanoma patient without a gallium scan. The thoracic surgeons are not yet using it routinely for their patients with bronchogenic carcinoma because I think it would take away 70 percent of their practice. I seriously believe that. Because in the patients—the bronchiogenics that we have received from the medical service before they are referred to the surgeons—we usually show bronchogenic carcinoma far beyond the level that is demonstrated by the standard work-up to include bronchoscopy, bronchonography, tomography, and everything else. So I personally think that all bronchiogenics should have a gallium-67 scan. You not only show the lesion, you show the extent of mediostinal disease and you may also see distant mestastases.

VERDON: I would like to echo that from our experience. One of the things that we have studied extensively has been the bronchogenic carcinoma and we have taken the approach that the study is being done to avoid thoracotomy. We have been particularly interested in looking at supraclavicular lymph node spread and other local sites of spread. We have had several cases now in which we have picked up nodes that were not palpable in the supraclavicular fossa and biopsy revealed positive nodes that range from 5 to 9 mm in size. These nodes supply the answer right there. Diagnosis and prognosis were determined with a simple surgical procedure. Now, when you showed some unusual tumors that picked up gallium, I would like to ask what should we be using gallium for? Should we be using gallium for everybody that comes in with every possible, conceivable tumor? Is that the role of gallium? Or should we be concentrating it more specifically on certain tumors. We have heard melanoma mentioned. We have heard bronchogenic carcinoma, lymphomas. Are there any other tumors that we can count on for predictability of gallium uptake?

WENZEL: I think that there are a lot of solid tumors that are not really well investigated at this point. I have a relatively small series of things that are highly positive in our hands. One is breast cancer, also carcinoma of the colon, and endometrial carcinomas. Amazing that this tumor would take up gallium. We have studied 15 cases of endometrial carcinoma and all have been positive. Pancreatic tumors do not pick up gallium to any predictable degree. Dr. Fordham, why don't you discuss your feelings on this question.

FORDHAM: We haven't looked at as many endometrials. They tend not to use the gallium scan on these patients until they are in some trouble. But I would say that you ought to be using the gallium scan in all your germinal cell tumors—testiculars and ovarians—not to evaluate the mass that can be felt, but to look for disease elsewhere. With respect to tumors of the gut, I would not tend to look at those, simply because if symptoms are related to the gut, they ought to go through the standard studies. I have no quarrel with doing a gallium scan but I don't think pickup rates are as high in adenocarcinomas, which is usually a gut lesion, as we see in lymphomas, melanomas, and so forth. There is another group of tumors, though, that I want very much to look at. We don't have extensive experience with these. They are the soft tissue sarcomas. We are finding out that we are detecting at least 75 percent of these unusual tumors. Our series is too small. But these strange birds are the ones I think should be looked at because it is tough to demonstrate a nice, clear, progression of disease pattern, and that is the very thing that you should use the gallium scan for. Because you wind up with a patient who has a lesion and you biopsy it and it is, say, a rhabdomyosarcoma. Where are you going to look for metastases—you do a chest x-ray and then where do you look? You really have nothing to go on unless you want to take the CT scanner and cut the patient from head to toe. And I don't think they will ever use CT scanners like that. They shouldn't. And I think that is exactly the role of the gallium scan. Particularly with the tomographic scanner which not only tells you the X and Y axis, where your disease is, but the Z axis as well.

VERDON: Would you take the approach, then, that we should be studying almost all tumor patients? Or should we break it down into certain groups?

FORDHAM: No, you are going to have to have about three or four tomographic scanners in your department in order to scan every patient with malignant disease. So I would tend to concentrate on the ones that we have already talked about.

VERDON: So we would agree that we should be studying lymphomas?

FORDHAM: Yes, all lymphomas, particularly histiocytic and Hodgkin's, all melanomas, germ cell tumors, and soft tissue sarcomas.

VERDON: What about hepatoma?

FORDAHM: I haven't seen enough to have a feeling for that.

WENZEL: Hepatoma is a pretty unusual tumor. But everyone that we have studied has been positive. My experience with the soft tissue tumors has not been very good, Dr. Fordham. We have studied several and our yield has been poor but, again, I am only talking in a realm of approximately 40 patients. Also, I feel that we should be studying tumors of childhood. We have had about 100 patients that we have studied and I think all tumors of childhood deserve a look with gallium because you get some pretty strange ones, especially the one that you brought up, Dr. Fordham, of rhabdomyosarcoma. Almost all of these lesions have been positive to gallium scans.

VERDON: Has anyone in the audience had any experience with gallium that they would like to add? We have not had luck with the GI tract tumors at all. Our yield has been down around 3 or 5 percent. And we have not really had very good luck with breast tumors. Has anyone had better luck with breasts? Since this is work primarily discussing the tomo scan, I would like to ask the two of you one question. Do you think you *frequently*—that is the key word—detect lesions with the tomo scanner that are not appreciated with other detection devices? Gallium, bone scan, liver scan, brain scan, any one you want. In other words, should an instituion buy a tomo scanner to back up their camera or whatever they are using now so that they are not missing lesions?

FORDHAM: No, I think if you compare the tomo scanner against a large field-of-view camera, you will see most of your gallium positive lesions with the camera. One of the problems is that you don't know where their depth is. And I disagree that in going to multiple projections you can sort things out, particularly when you have a number of lesions. I would never buy a tomo scanner in place of a large field-of-view camera. You have to have one of those in your laboratory first. Maybe you ought to have two in your laboratory, but I think if you buy three or four cameras, you are crazy. You should buy a tomo scanner, which gives you a whole different perspective. I really do think that if you are doing a significant amount of cancer work or a significant amount of gallium work that the tomographic scanner is a must because again, it gives you depth, as well as distribution.

The answer to the other question alluded to is yes, we do see some lesions that are detected on a tomo scanner that are not appreciated on the scintillation camera. It is a small percentage, but we do see them. When we compare our experience to the rectilinear scanner for anything, the tomographic scanner will beat it hands down across the board. As far as I am concerned, the rectilinear scanner is dead. The tomographic scanner will do everything a standard rectilinear scanner will do and do it far superiorly.

VERDON: How long does it take to do a tomographic scan?

FORDHAM: The same length of time it takes to do it with any other device. Gallium scans take about an hour. On our gallium scans, we image from the head to below the hips, we don't go below the hips unless there is a problem in the leg. This study takes about an hour. For standard bone scans, we run the machine faster and study the whole body. This again takes approximately an hour. You get all 12 planes at the same time and you don't have to scan the patient 12 times.

VERDON: Dr. Wenzel, would you like to add anything to that?

WENZEL: No.

VERDON: I have another question here from the audience. This is a jump ball so whoever gets it can handle it. How frequently do you see negative bone scans but positive bone surveys for metastatic cancer? In other words, what is our rate of false negatives with bone scans?

WENZEL: I would guess less than 1 percent.

FORDHAM: I would put it higher, maybe as high as 2 percent, no more than that. The ones you miss—you know, Charkes is smarter than the whole bunch of us. Back in 1968, he talked about 5 percent false negative rate. And he was right because about 3 percent of the lesions will show only in the peripheral skeleton and in those days of strontium-85, nobody was scanning the peripheral skeleton. But if you go back then, he has about the same 2 percent that we should be having now. And the ones you miss are the same ones we have always missed. It is the roentgenographically osteoblastic disease. Why? Because the disease is indolent, the body has a lot of time and lays down bone in a very slow, easy-going fashion. And on a bone scan you are not looking at what has happened, you are looking at what is happening now. So the lesions that you see are the lytic lesions where you are having an immediate reaction to the invading tumor. A second category that you miss is that axial peripheral disproportion I was telling you about. You shouldn't be missing them. The third area that we tend to miss are groups of tumors that do not arouse a response by bones, such as myeloma. Most of the tumors that we have trouble with are the ones that look like marrow components, myeloma, lymphoma, and even oat cell carcinoma. The osteoblastic and osteoclastic responses usually seen as bone response to tumor infiltration is lacking in the above-named lesions.

VERDON: Since you have nicely described that, Dr. Fordham, could you now tell us why we do get a positive bone scan? What is the current thinking on the mechanism as to why a bone scan is positive?

FORDHAM: Well, I think it is related to every place where you see osseous remodeling, osseous rebuilding, and that happens in normal bone, too, but at a reduced rate. There are many mechanisms reported, access to bone crystal, blood flow, and everything else. But nonetheless, I think what you see on the bone scan actively reflects what is happening with those osteoblasts and those osteoclasts microscopically. As to chemically, what has happened, I just don't know.

VERDON: What we do in any case of myeloma is recommend skeletal survey in addition to the bone scan and we half jokingly say that the scan has a 50 percent false positive and a 50 percent false negative rate in evaluation of this disease. This is very true. We frequently see x-rays that are totally involved with tumor and yet we obtain a negative bone scan on a patient with myeloma. But we have also seen patients with negative x-rays and a markedly abnormal bone scan in a patient with the same disease. With myeloma, you have to pull out all the diagnostic modalities you have.

Here is a question that someone from the audience has asked. I would like for you to comment on this, too. Please comment concerning the performance of bowel prep prior to gallium imaging. This is one of the hottest issues we have in our hospital. If one reviews the literature, one can surmise that there is no evidence that the activity that we see on the scan is solely in the stool and there is good evidence that there is mucosal activity which would not change, no matter how much prepping we do. And third, many have prepped patients until they begged for mercy with no clearing of gallium from the colon. This is a very good question and raises a very practical point. I think that the biggest problem that we have with gallium scanning today is that the patients don't mind being studied but they say, "Don't give us that bowel prep."

WENZEL: Well, we give them the bowel prep anyway. And the main reason is that most of the time, although we have had a sort of mucosal hang-up in the normal mucosa, it usually comes out with a reasonable prep. Most of the patients don't complain, but if they do I try a little switch on the second day—something milder, or I just sometimes will have them come back a second day, or I have even gone in and palpated the abdomen and tried to move things around a bit. Quite honestly, especially in children, if you want to get an answer more quickly, see if you can't stimulate some kind of movement.

FORDHAM: We are trying something that appears to be working. It dawned on me that I don't like castor oil. And it dawned on me, after we had a particularly troublesome study where we had a patient and scanned him many times. We usually scan at 72 hours. On this man we scanned at 96 hours and prior to each rescan we administered castor oil. On this individual, we scanned three or four times and administered castor oil to him on each occasion. I talked to him, thinking that maybe he wasn't taking his castor oil and I found the poor man was just utterly drained. He said, "Doc, what are you trying to do? How can I poop if there ain't nothing to poop?" And I said, "What's the matter?" And he said, "I am on a liquid diet." So bells rang. So you know what we do now and it works beautifully. Approximately 90 percent of the time, we get excellent results. We don't give the patients castor oil, we don't give them anything, but what we have them do for three or four days is to eat no rice, no potatoes. We have them eat high fiber bran bread and bran flakes for breakfast every morning. We give them salads with every meal and fruit until it comes out their ears. We find this really works. I think one of the problems is that you have to move bulk through. You don't want the colon empty for a gallium scan. This is not the same as for a barium enema. All you want to do is have the material that's collected there in the first two or three days move on through, and the best way to move it on through is to give them a diet that will help them move it. We find this makes the patients much happier. We are using this procedure routinely on our outpatients and we find that we only have to rescan patients at a rate of approximately one out of ten. For inpatients, it is troublesome because there are so many problems with diet and everything else in the house, so many other tests that they are getting pressed for. Matter of fact, I think that the castor oil may induce an inflammatory response in the colon that actually increases the uptake of gallium. I have had that nasty feeling on occasion, but I cannot prove it.

VERDON: Dr. Fordham, what is your routine? Do you do one gallium scan or do you do them periodically for several days?

FORDHAM: As a rule, we scan them once, we inject them Thursday for a Monday scan, Friday for a Tuesday scan, Monday for a Thursday, and Tuesday for a Friday. In other words, 72 hours apart. We keep Wednesdays open for particular things such as a 96-hour scan or other rescans.

VERDON: We have a little different philosophy at our institution. We start our tumor scans at 48 hours and carry them through to 72 hours, and then if there is a questionable area, particularly in the chest, we will zero in on that area and study only it in a 96-hour study. And in some cases, not jokingly, we won't quit until the last photon has been fired. Now how about for abscess scanning, we know the literature says if you look at 2 hours or maybe 3, you can diagnose an abscess immediately. Do you have any different time sequence with gallium? Will you accept just a 2-hour scan and quit, or will you go further?

FORDHAM: We use the same timing if we can keep the surgeons off our backs. Occasionally, I will submit to a 48-hour scan but I find that if you talk to the surgeons and reason with them, they will usually wait longer and so we just stay with a 72-hour scan.

VERDON: Does anyone in the audience have a comment concerning an abscess search assuming that we do have time to study the patients? Is there any value in doing a 2-, 4-, or 6-hour scan?

FORDHAM: We will occasionally go with a 6-hour scan. I will not go with a 24-hour scan because I think that is the worst time to start scanning. So we go ahead with a 6-hour scan occasionally. The point is that if we had the opportunity to look at everybody with inflammatory disease, I would love to do 6-hour scans. But the kind of patients we study have been around for weeks, usually with an unusual clinical situation or fever of unknown origin or anything else. They have been sitting around for a few weeks and my philosophy is that if they waited that long, they can wait 3 days more. If they are acute, you can't keep the surgeon out of the abdomen and if they are not that acute, then there is time to wait. Even if we do a 6-hour scan and it is positive, we will rarely go on that as our final diagnosis. Surgeons will usually wait and we do get the 72-hour scan to see it better. So in general, we prefer the 72-hour scan for abscess studies as well.

VERDON: Does anyone in the audience have any comments?

DR. HANDMAKER (IN THE AUDIENCE): In cases of questionable osteomyelitis, you would want to know as soon as possible so that the patient could be started on antibiotics promptly. We have started doing all our gallium scans for osteomyelitis at 3 to 6 hours. In the extremity, you don't have as much problem as you do in the abdomen because there is not a problem with bowel activity.

VERDON: Good point, Dr. Handmaker. Another question for Dr. Fordham. Radiographically, one may diagnose hyperparathyroidism of bone by preferential location of bone erosion—radial aspect, medial phalanges, clavicle, etc. Is there any correlation between the isotope and radiographic lesions?

FORDHAM: Yes, I think there is. I think it is the

same old story, though; you will see the disease sooner on bone scan than you will see it radiographically. Therefore, the pattern will be different if it's being seen before it has reached a stage that it can be diagnosed by conventional x-ray.

VERDON: Does anyone in the audience have any particular questions they would like to present at this time?

QUESTION: I would like to know the false positive and false negative rate of gallium scans.

VERDON: In any specific disease state or just across the board use of gallium?

COMMENT: Across the board use.

WENZEL: Dr. Handmaker had a very nice article with Dr. O'Mara in which he reviewed the situation. Our experience with this situation has been the same in a series that we have done but have not published. The biggest problem with false positives is in inflammatory lesions. The problem with false positives occurs usually in the abdomen and this is due to bowel concentration of gallium. False negatives below the diaphragm in our series was around 15 percent. Overall positivity of detecting tumors which were proven by biopsy was 76 percent in all tumors.

VERDON: Any other questions?

QUESTION: I was wondering about the cost of the gallium scan as compared with the radiology procedure?

FORDHAM: Let me say that anytime you do whole body imaging with the radionuclide method, it is going to be far cheaper than anything you do roentgenographically. You can't cover the skeleton roentgenographically for under hundreds of dollars, where you would have to take a survey of this bone and a picture of that bone and certain special views of various bones. But you are usually looking at them with one view at a time which is a very dangerous practice. The actual cost is always far greater roentgenographically because I defy you to find me a radiologist who can look at a roentgenographic survey with anywhere near the accuracy that I can and in anything less than about 50 times as much of his professional time. You put up the bone scan, you put up the gallium scan; in 2 to 5 seconds, you usually decide whether it is normal or abnormal. This cannot be done on a metastatic survey in anywhere near this time frame.

VERDON: What is the cost of a bone scan and a gallium scan on the Pho/Con?

FORDHAM: The Pho/Con really costs no more than a large field-of-view camera with a moving table. They are very comparable. So equipment costs should not be any greater. We charge $175 for a bone scan and we charge about $200 for a gallium scan.

Ultrasound in Gynecology

Edward A. Lyons

Ultrasound plays a major role in the examination of the pelvis of the pregnant and nonpregnant female. Using the full bladder technique, I can readily visualize the normal vagina, uterus, tubes, and ovaries. With careful examination, lesions in the order of 0.5 cm can be identified within the uterus or ovaries. This chapter will strive to point out normal anatomical features in the female pelvis and then to highlight some variations of normal, as well as some of the common pathological conditions likely to be encountered in everyday practice.

The size of the uterus and the character of the central cavity echo may change in response to normal as well as pathological states. A premenstrual uterus will have a thick proliferative endometrium with a bright central cavity echo, whereas a menstruating uterus may have an irregular, relatively echo-free central cavity. Very intense central echoes may be produced by intrauterine contraceptive devices (IUCD). This is one field in which ultrasound can make a tremendous contribution. In the nongravid uterus one should always be able to identify an IUCD. This is also true in the early gravid uterus. One can miss an IUCD if it is outside the uterus or if it lies within a large gravid uterus. The specific features to aid one in identifying an IUCD are: (1) a highly echogenic structure, and (2) a double line or series of double lines about 2.5 mm apart. This represents the front and back walls of the IUCD. In order to clearly identify this, one should magnify the image so that the entire uterus fills the screen.

Uterine fibroids are a common condition. One of the most useful identifying features is the remarkably poor penetration exhibited at high frequencies with marked improvement at lower frequencies. At 3.5 Mhz, an undegenerated fibroid will appear almost echo-free with a very poor back wall. At 2.25 Mhz a good back and considerable number of echoes will be identified. Once the mass begins to degenerate, more echoes are seen, even at high frequencies. Calcification within a fibroid will also produce strong echoes.

Pelvic inflammatory disease may have some confusing ultrasonic appearances. Generally, if one follows the patient over a 2-week period some characteristic features will become evident. It is important to realize that this can be confused with an ectopic pregnancy. Initially we see a swelling of the fallopian tube, often measuring greater than 1 cm in AP diameter and being less echogenic than normal. This may be associated with a variety of oblong fluid collections in and around the uterus. This may represent fluid in the tubes as well as in the bowel. Later on a distinct collection in the tube, which may have varying amounts of echoes within it, will be seen. This may all resolve or may go on to become a large oblong relatively echo-free mass which represents a hydrosalpinx.

Ovarian cysts should not pose a major diagnostic problem. They are generally rounded, well defined, and echo-free and have a good back wall and good through transmission. They may be septated or may have a very thick wall if they become infarcted. The finding of solid components within a cystic mass or finding of debris may indicate the presence of a dermoid or of a papillary cystadenocarcinoma. It is often difficult to tell the difference between mucinous cystadenoma and a serous cystadenoma. Endometriomas may present with a variety of ultrasonic features. The mass is most often somewhat irregular in outline and may be echo-free or may have echoes within it. When combined with the history, the diagnosis is readily made.

Ovarian tumors often have complex appearances with both solid and cystic components arranged in an erratic fashion. Ascites is often a dominant feature. Regardless of the patient's age, a complex pelvic mass with ascites is an ominous finding.

Discussion: Obstetric and Gynecologic Imaging

Moderator: Thomas A. Verdon
Panelists: Edward A. Lyons
Fred Winsberg

VERDON: In your presentations, you both placed a lot of stress on the pregnancy test in conjunction with your interpretation. Whether the test was positive or negative played a major role in the interpretation of the ultrasound procedure. Do either of you have any feelings as to which pregnancy test appears to be the best? There are many new pregnancy tests on the market, including one that will enable you to detect pregnancy as soon as 8 days postconception. Do you have any experience with using these tests?

WINSBERG: The best test is the test of the beta sub unit. The problem is that at our hospital, we only perform it once a week. That is usually not fast enough to help you with a particular clinical problem. I have found that a situation such as the ectopic pregnancy, where you would really like to rely on a pregnancy test, you frequently can't get it or what you do get is unreliable. So now, I don't place a lot of emphasis on a pregnancy test. Now, that may be a deficiency of our hospital in that they don't do them well or they don't do them frequently enough, and other people who have better laboratory services may not have that problem.

LYONS: Just to mention one point, in our lab, they do the Gravindex pregnancy test which uses urine and relies on the HCG and you get agglutination with some various fluids that they provide with the test. However, one point with that particular test, and I don't know if any of you use it, is that the test remains positive for up to 3 days following a complete abortion. So, you may have a woman who has completely expelled all the products of conception and comes in 2 days later with a positive pregnancy test and an empty uterus. Then you don't know where you are. Let me just make one more thing clear. I think that you should always give the fetus the benefit of the doubt. We never make a diagnosis of blighted ovum or fetal death, for instance, on one examination, unless, of course, it is very apparent. More often than not, we have had problems in interpretation. The woman's history is usually very, very poor and even in these women who come in and say they have been pregnant for 12 weeks and you see a tiny little fetal sac, nothing is stopping the first fetus from aborting completely and this woman from getting pregnant a second time. That has happened to us morethan once. We routinely do a second examination 1 week later on cases of fetal death. It doesn't cost anything extra in time and that way you will not abort a normal fetus. No one is going to thank you for not aborting a normal fetus. It only costs you 1 week in time and I think as a general rule, always repeat the cases of blighted ovum any time you are in doubt. Blighted ovum, fetal death, or whatever term you choose to use; always repeat them in about 1 week's time.

VERDON: That's true, Dr. Lyons, that no one is going to thank you for not aborting a normal fetus, but you find that you do sleep a little better at night using that approach.

WINSBERG: The situation that Dr. Lyons is talking about is the patient who comes in with a history of, say, 12 weeks of amenorrhea. He is quite right. The patient might have a history of 12 weeks of amenorrhea or 18 weeks of amenorrhea but really only have a 5-week gestation. I think it depends, really, in that situation, on how reliable you think the history is and how much that mother really wants to keep the fetus. That is a situation which is not occurring as often today as it used to. Another problem is that if you are conservative and decide to wait a week, that patient might come in in the middle of the night with a spontaneous abortion and a serious bleeding problem. So you have to balance it one way or the other. It is a touchy problem. Some of the uteri are so obviously abnormal in their size and shape that it is almost inconceivable, if I can use a bad pun, that there would be a pregnancy in that uterus.

LYONS: Sometimes you can get caught on that because an overfull bladder can flatten out that uterus, and you may get odd shapes of the uterus. We still try to repeat all these cases in approximately 1 week.

VERDON: If the placenta is as compressible as it seems, why would not the results of your pushing on the fetal head to determine the presence or absence of the posterior low-lying placenta be the opposite to that suggested?

WINSBERG: I understand the question. The placenta is somewhat compressible. But if you have a placenta sitting between the fetal head and the sacrum, you cannot push that fetal head all the way down to the sacrum. This is an old observation that was noted in classical radiology even before the days of ultrasound. As I indicated to you, another maneuver which is equally valid and perhaps better is to put the patient in a Trendelenberg position and try to push the head up out of the way and actually positively identify the location of the placenta. But with the placenta, a posterior placenta, sitting on top of the cervix, you will alwayshave at least a 1- or 2-cm space that will be between the sacrum and the fetal head. This is the chorionic plate and it is somewhat compressible. That is kind of surprising, when you see this phenomenon in real-time, you see the fetal hands moving around, squeezing against the chorionic plate, and it is compressible.

VERDON: I would like to raise a question now. We don't like to do anything routinely in the practice of medicine. Right away everyone gets upset if you mention performing a routine procedure. But what would be wrong with performing an ultrasound examination on every intrauterine pregnancy at a reasonable point in time, i.e., second trimester? We all know conception dates are frequently inaccurate. You could tell more accurately the actual dates and age of the fetus, you could possibly tell the presence and absence of fetal anomalies, and with serial studies you could determine the presence of intrauterine growth retardation. Do you see anything wrong with this approach other than it being a routine procedure?

WINSBERG: Well, we are still not 100 percent sure that there are no deliterious effects to the use of ultrasound. I think we are 99.9 percent sure, but we are not 100 percent sure. Other than that, I think the only contraindication to that approach is the absence of enough machines and people to perform the number of studies required. I think that as the machines develop, which they certainly will do over the next few years, and we have really high quality real-time machines such that we can perform a thorough examination within a very short period of time, I certainly can conceive of this as something that will come, unless someone can show that there are some adverse biological effects about which we do not yet know.

LYONS: Just when you said that, Dr. Winsberg, I was thinking of the little x-ray machines that they used to have in shoe stores where every little kid used to spend the afternoon watching his toes. Develop a really good real-time machine and a woman just goes up to it and looks at the fetus. No, I do not think it should be used routinely. First of all, we are not 100 percent sure that ultrasound is absolutely harmless, and there is some recent work that shows that even using the power levels that we are using, you are getting some noticable effects in lymphocytes. Granted, this was done in vitro, in the test tube. But this is the first time that anyone has shown any effects. So, in fact, maybe there is something happening and maybe we have just always been looking at the wrong system. It is my philosophy that if you don't need the tests—if the woman doesn't need the test—she shouldn't have it. You shouldn't do these things unnecessarily. This has been suggested in Britain, but again, there are just not enough machines or technologists to study all the people who are pregnant.

WINSBERG: One anencephalic that I showed during my presentation was detected on a study that was done for no apparent reason. Also, we occasionally pick up twin pregnancies which are not suspected. Obviously, the percentage of things that you are going to pick up just examining every pregnancy is relatively small, But we are now at a stage of zero population growth and I think people are interested in the quality of their offspring and it seems to me that the trend will be, 10 or 15 years from now, that every pregnancy will be examined by ultrasound, unless somebody can show that it is harmful.

VERDON: Dr. Lyons, another problem in use of ultrasound is in the differentiation of cystic versus solid masses. You stated you never pay attention to the internal echoes. I think that is a little strong. Would you care to explain this point?

LYONS: What I meant was never try and establish whether or not a structure is solid or cystic by what its internal characteristics are. Never begin at that point. Certainly there is important information to be received by examining the internal echoes of various structures. As I told you, with fibroids, you get poor back wall transmission and very few echoes within the fibroid tumor. That is diagnostic. Also in lymphomas, again, we find solid masses with very, very few internal echoes. Again, there are diagnostic features that we have been able to recognize. But never begin by looking at the internal echoes to decide whether or not the abnormality is cystic or solid.

WINSBERG: May I make the suggestion that we stop using the terms cystic and solid and instead use the terms solid and liquid. I think we probably would be more accurate. We are not really talking about cystic versus solid masses. We are talking about solid masses versus liquid masses. You can have part of a liquid mass that has some internal echoes. You have to keep several things in mind. You have several factors playing a role. One is attenuation. The liquid essentially does not attenuate the sound and therefore the sound goes through it and you get posterior wall image which is very strong. On the other hand, you can have structures which have very little in the way of internal echoes which attenuate sounds, such as lymphomas or fibroids. And here you get poor posterior wall reinforcement. So, be careful of trying to separate liquid from solid masses purely on an internal echo pattern.

The Role of Radionuclide Liver Scanning

Wayne W. Wenzel

There are two categories of liver radiopharmaceuticals currently in use (Table 1). The polygonal cells make up 85 percent of the total liver bulk and are the functional cells of the liver. Functional liver agents have been used in the past and are still used to assess liver function. The Kupffer cells make up only 15 percent of the liver mass, but these cells are the ones most frequently imaged and will be the topic of this discussion.

Technetium sulfur colloid is currently the agent of choice for isotope liver imaging and will be the agent that was used for all isotope liver scans seen in this chapter.

When gold colloid was the liver imaging agent, we were lucky to be able to do two views in a reasonable amount of time. Currently, most people do a minimum of six views and additional obliques as needed (Fig. 1). On one of the views, most commonly the anterior view, most laboratories will place anatomical lead markers to give them some sort of reference. The one we use is lead markers at the costal margin plus a lead rectangle which measures 10 cm from center to center on the long axis and 5 cm from center to center on the short axis. Using this design, one cannot be misled by the image being too dark and ballooning effect causing foreshortening of the projected image. We feel that some sort of image-measuring device is necessary for liver imaging and that the upper limits of normal of

Table 1. Liver radiopharmaceuticals

Polygonal (85%)	*Kupffer (15%)*
I-131 rose Bengal	Au-198 colloid
Tc-99m Pyg	In-113m hydroxide
Tc-99m DHTA	Tc-99m S-colloid
Tc-99m Diethyl IDA	

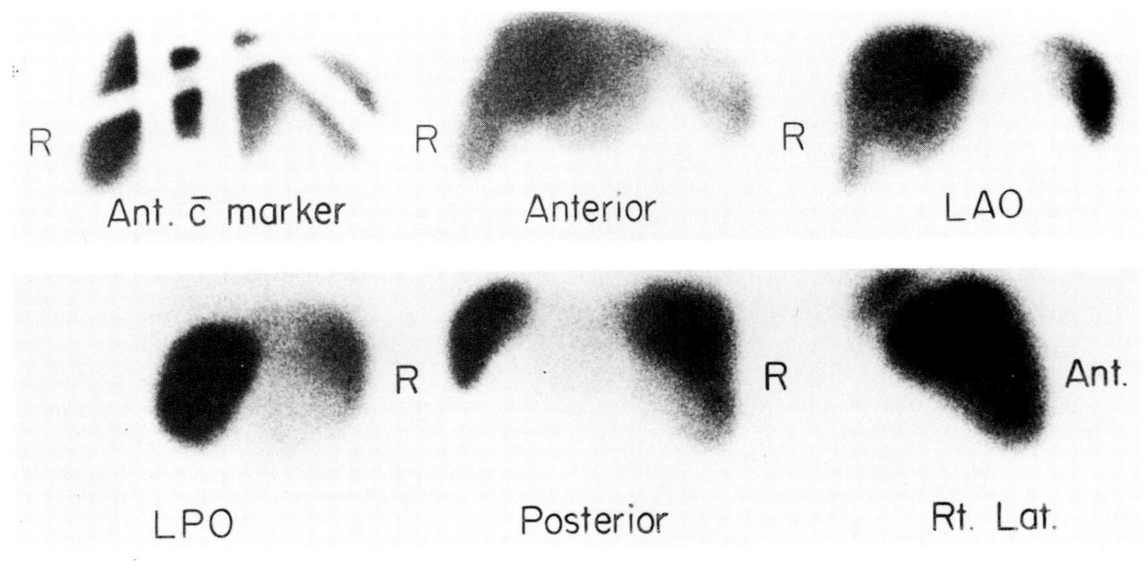

Figure 1. Normal liver.

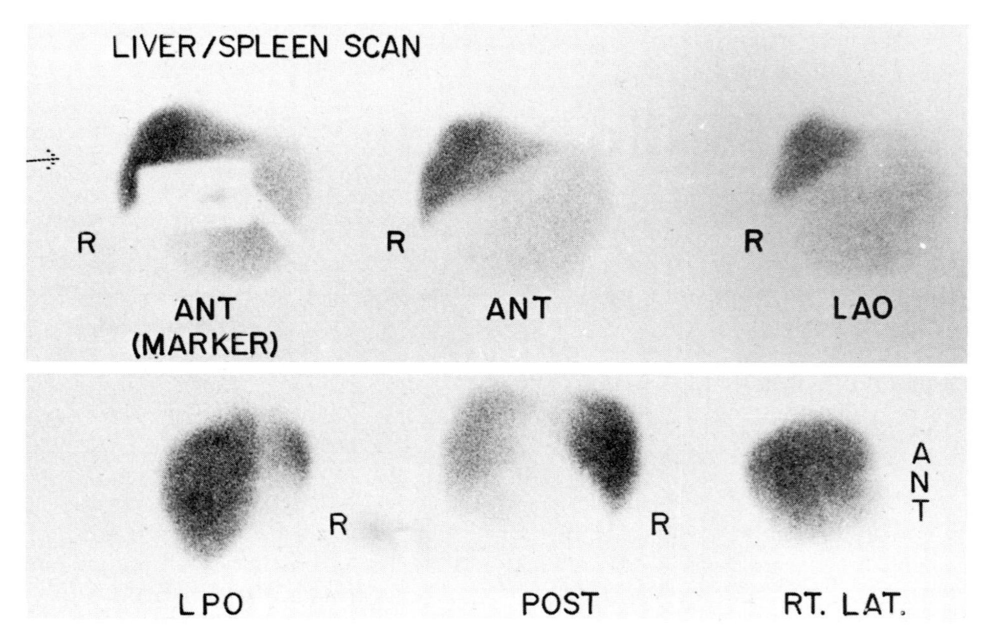

Figure 2. Splenomegaly secondary to leukemia.

Table 2. Indications for nuclide liver imaging

Liver size and shape
Primary or secondary neoplasm
Abscess or cyst
Cirrhosis
Trauma
Liver function

the cephalocaudal height of the liver is 15 cm. The upper limits of normal of the spleen is approximately 13 cm in cephalocaudal height.

Other devices which have improved overall resolution of liver images include large field-of-view cameras with high resolution parallel hole collimators and motion correction devices to minimize the effect of diaphragmatic motion.

Table 2 outlines the indications for radionuclide liver imaging and will serve as a guideline for this discussion. Perhaps the strongest argument for nuclide liver imaging is the ease and accuracy of liver and spleen size and shape determination.

Figure 2 demonstrates an example of a patient with massive splenomegaly with acute leukemia. To be sure, one can fairly accurately assess the spleen size with ultrasound or CT imaging, but radionuclide imaging is most efficient for this information. In addition (Fig 3.), one can easily, with a radioactive marker, place marks on the skin to outline the spleen so that the radiation oncologist can accurately direct his treatment beam.

Location of the spleen is not a frequent clinical problem, but Figure 4 shows an anterior liver scan on a patient with right lower quadrant pain which turned out to be a displaced spleen in the right lower quadrant on an exceptionally long pedicle coming from the left upper quadrant to the right lower quadrant. The long pedicle had undergone torsion and was causing the right lower quadrant pain.

Filling defects in the spleen seen on liver–spleen scanning are not common (Fig 5). Figure 5 shows an anterior and LPO view on a liver–spleen scan showing two distinct filling defects in the spleen. The differential diagnosis in such a case would include: trauma; infarction of unknown etiology; inflammation with abscess formation; cyst; or metastases from lymphoma or melanoma. In this case, there was a gallium scan performed which showed bright uptake in both of these filling defects consistent with neoplasm which, in this case, was lymphoma.

Liver size, as in the patient in Figure 6 with rather marked hepatomegaly, is most easily and accurately evaluated with nuclide liver imaging. This patient's hepatomegaly happened to be secondary to aspergillosis.

Liver position is frequently a clinical problem as the liver can often present as a palpable tender mass almost anywhere in the abdomen, and nuclide imaging is the easiest way to rule out normal variation of liver.

Figure 7 shows a chest x-ray of a child with an anterior mediastinal mass. A sharp pediatrician asked for a liver–spleen scan before exploratory surgery which showed that the mass was merely herniated liver

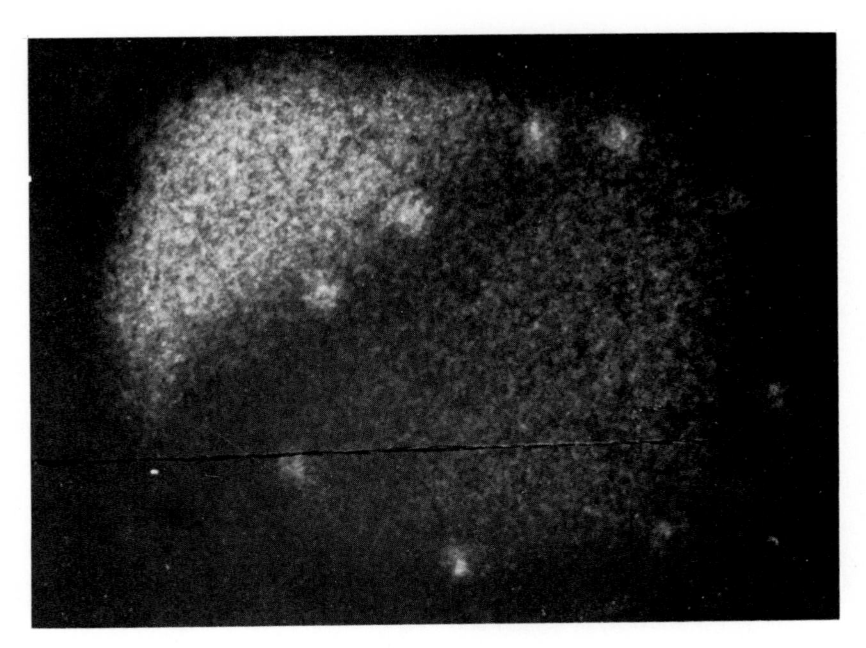

Figure 3. Anterior view with radioactive markers over skin marks over spleen. View done to verify radiation portal for spleen treatment.

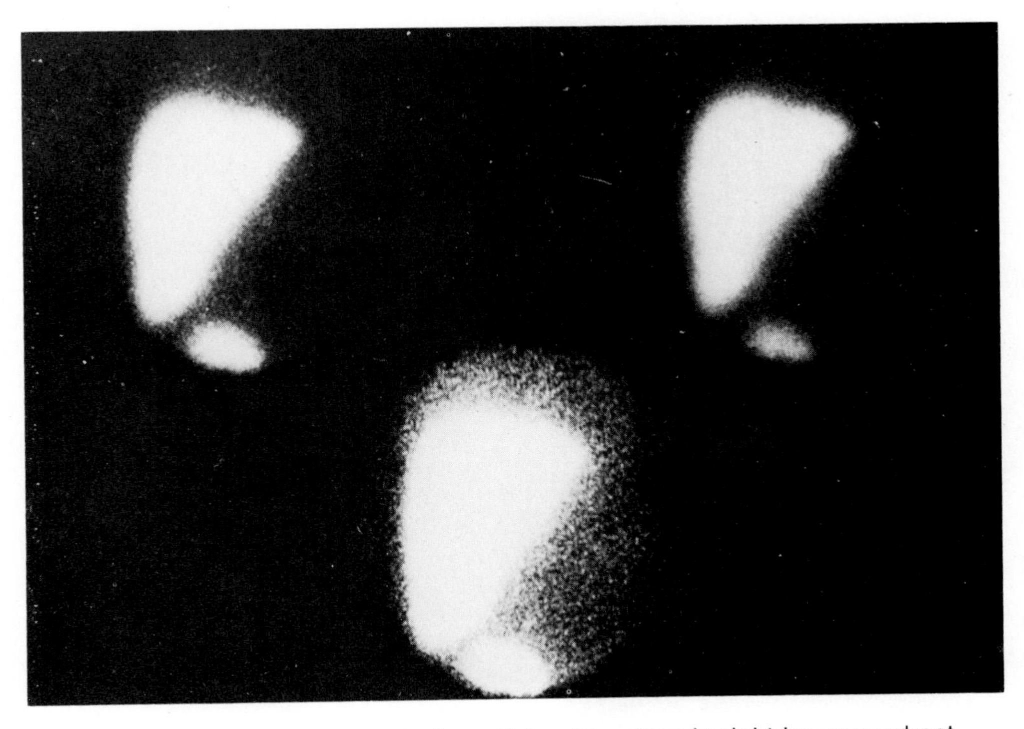

Figure 4. Torsion of long splenic pedicle with spleen in right lower quadrant.

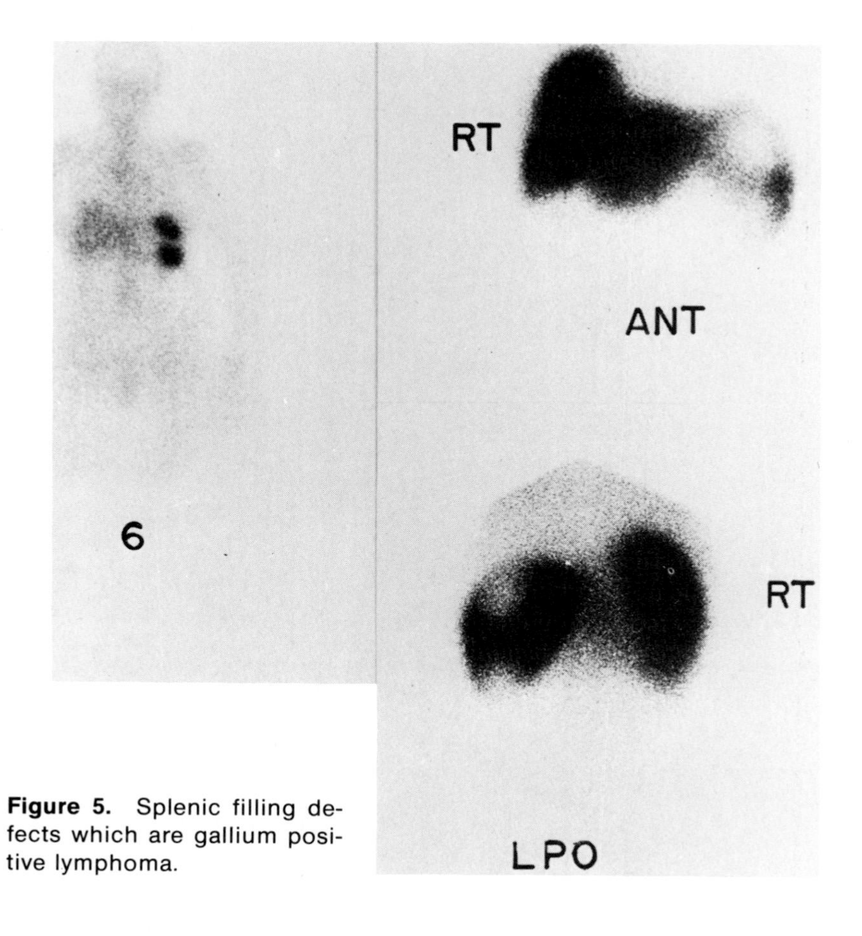

Figure 5. Splenic filling defects which are gallium positive lymphoma.

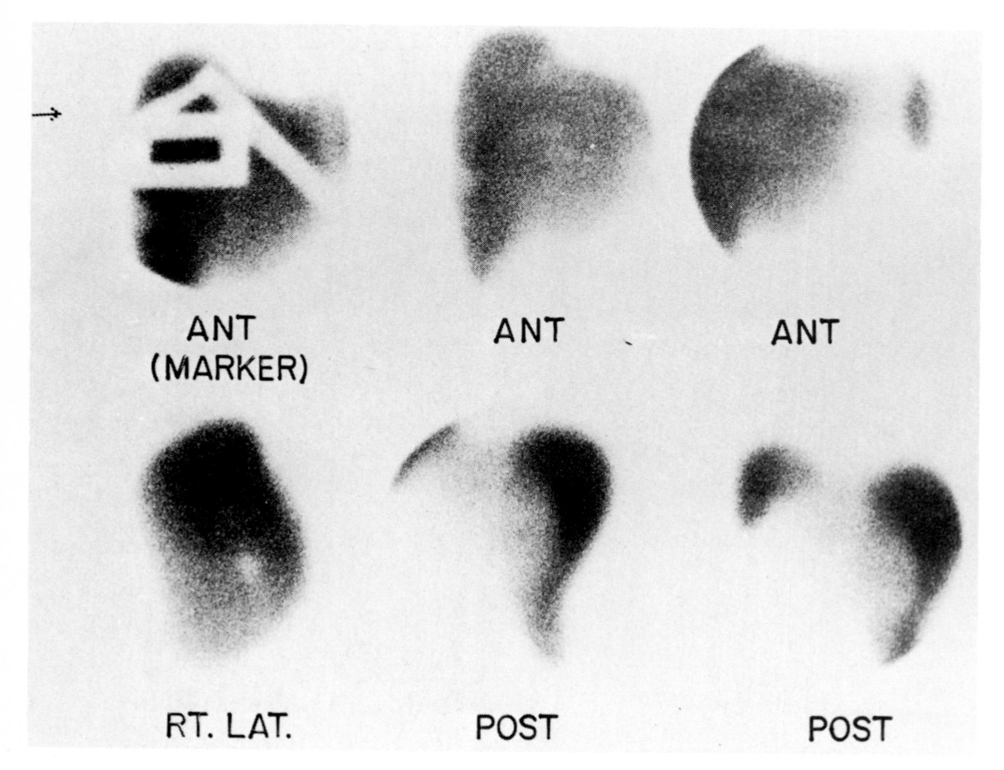

Figure 6. Hepatomegaly secondary to aspergillosis.

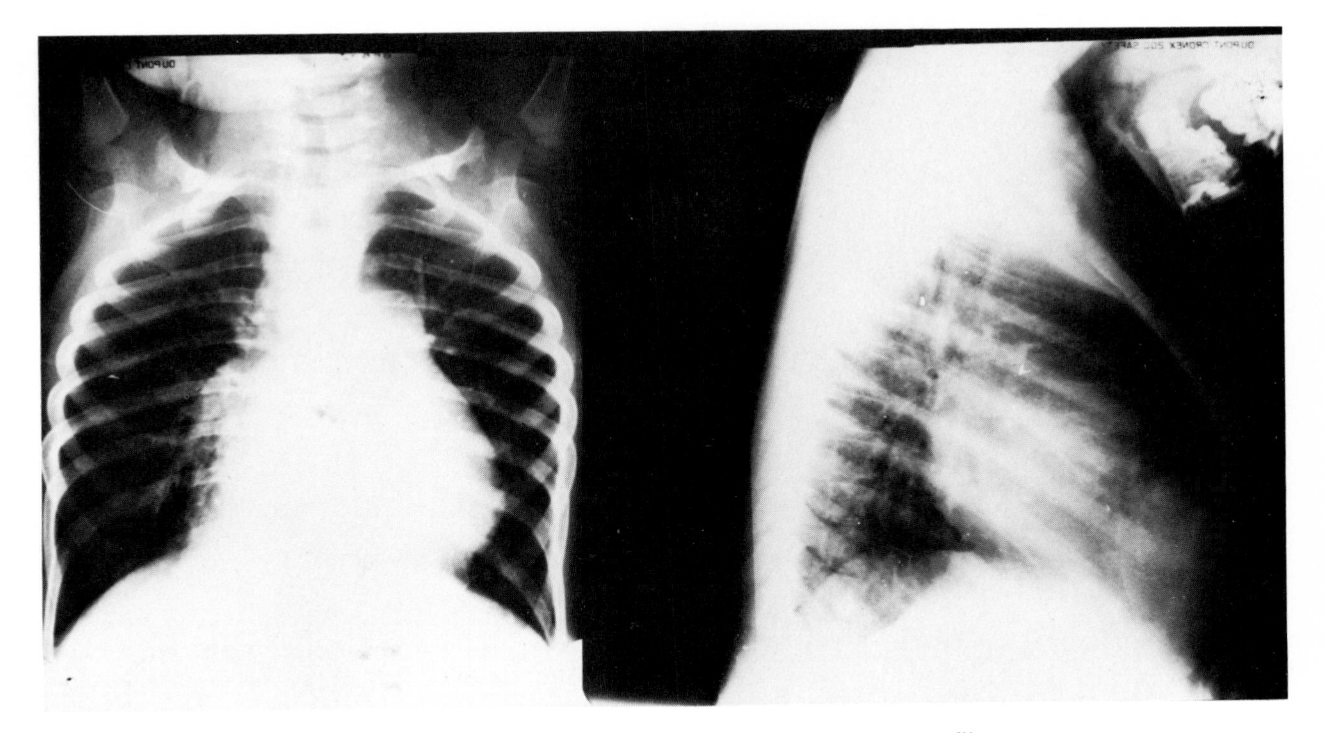

Figure 7. Mediastinal mass seen best on lateral chest film.

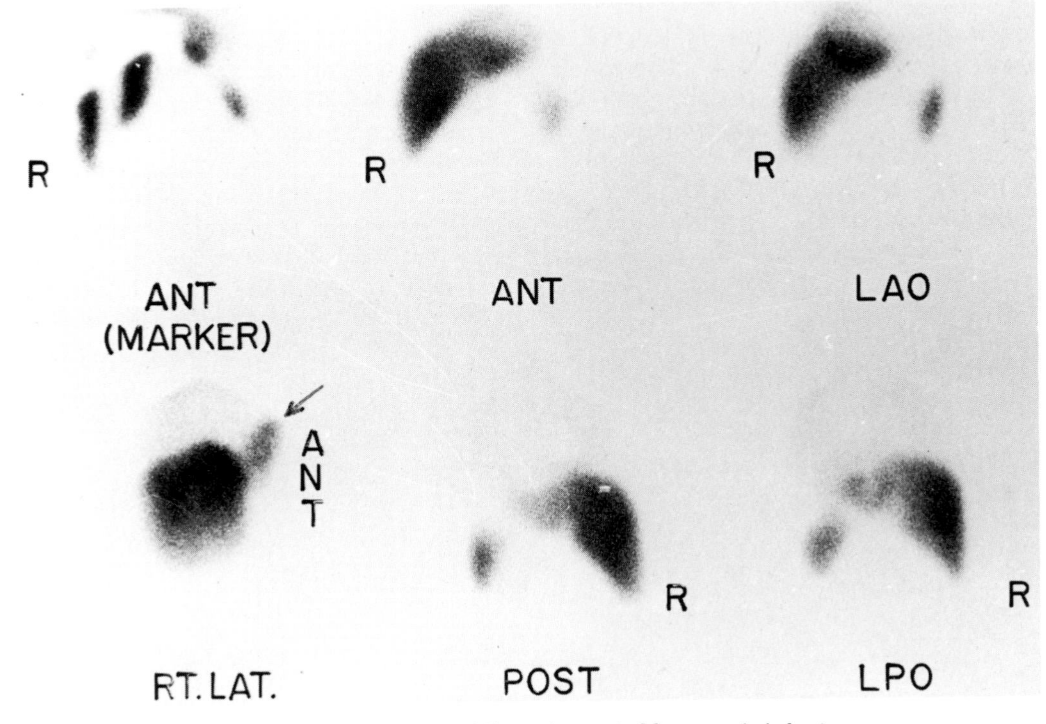

Figure 8. Herniated liver through Morgagni defect.

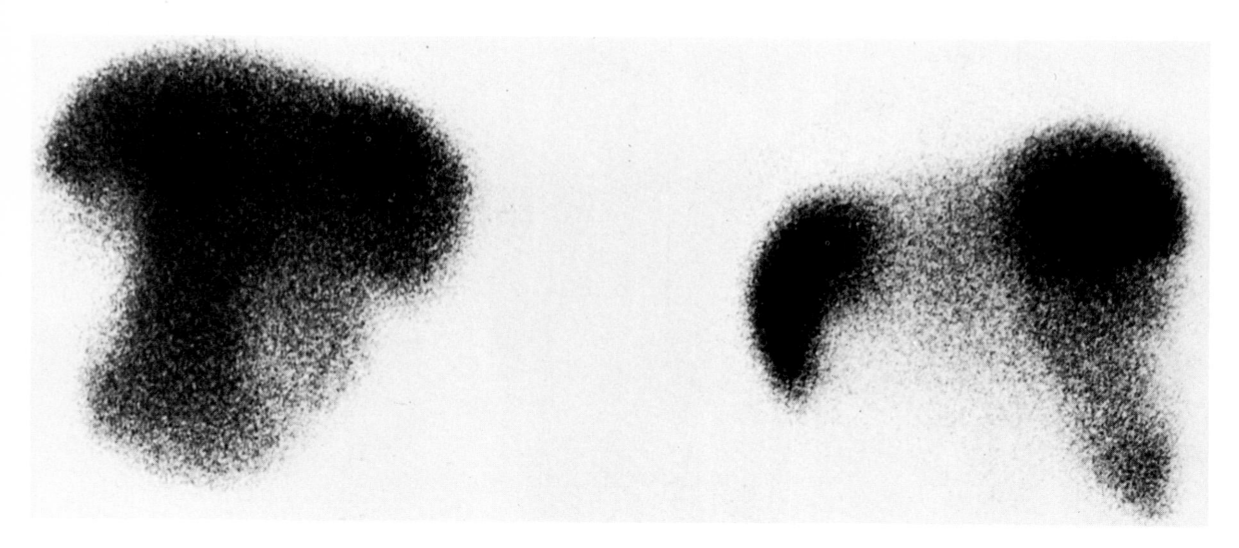

Figure 9. Liver defect secondary to subcapsular hematoma.

through an anterior diaphragmatic hernia of Morgagni (Fig 8).

Figure 9 shows a patient who was hit on the right side by falling rocks while working on a construction team in the mountains. A search for liver trauma can be done with ultrasound or CT scanning, but nuclide scanning affords a less expensive global look at the liver with less radiation exposure than CT scanning, and obviating the need for physical contact with the patient which is necessary with ultrasound scanning. Trauma patients frequently will have fractures and tender abrasions immediately adjacent to the liver, making ultrasound evaluation for liver trauma difficult. It is not necessary to touch the patient while doing nuclide imaging, which is a definite advantage in trauma patients.

Search for liver metastases remains the most common indication for nuclide liver imaging, and Figure 10 shows a right lateral view on two different dates of a patient with liver metastases secondary to carcinoma of the colon. It is in this group of patients where the addition of ultrasound imaging of the liver affords greater specificity to the nuclear image. Figure 11 demonstrates this point in that the nuclide liver scan seems to show only one large lesion in the liver and the surgeon was contemplating partial hepatectomy in this case. However, ultrasound evaluation showed that in addition to the single large lesion in the center of the liver, there were multiple satellite lesions posterior to it, making surgery impossible. Figure 11 demonstrates one of the four recognized patterns of liver metastases, namely the "bull's eye" pattern with a central dense echo surrounded by a clear halo, which is presumably edema.

Figure 12 shows an example of the so-called "mixed pattern" of metastatic disease, which is basically a disarray of the normal homogeneous echo pattern expected throughout the liver. Again, the nuclide liver scan shows an abnormality, but is nonspecific with regard to the etiology. The addition of ultrasound rules out the possibility of cystic masses in this patient.

Figure 13 shows the third ultrasonic pattern of hepatic metastatic disease, which is the "focal dense echo

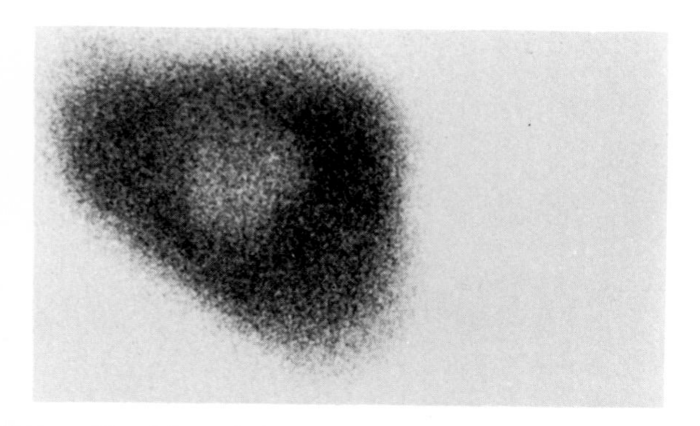

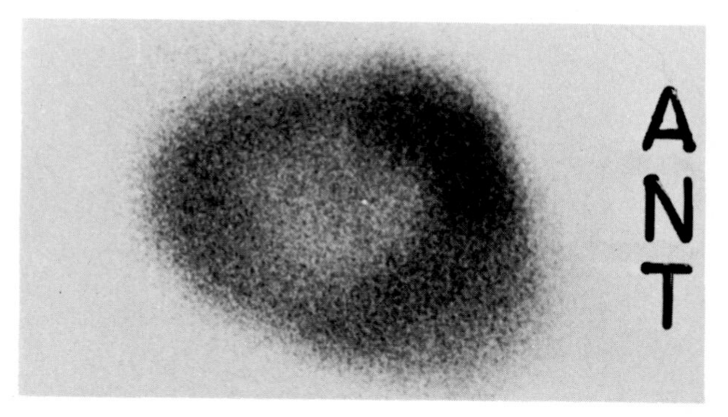

Figure 10. Liver defect showing growth over 4-month period. Left: Right lateral, 2/7/77. Right: Right lateral, 6/7/77.

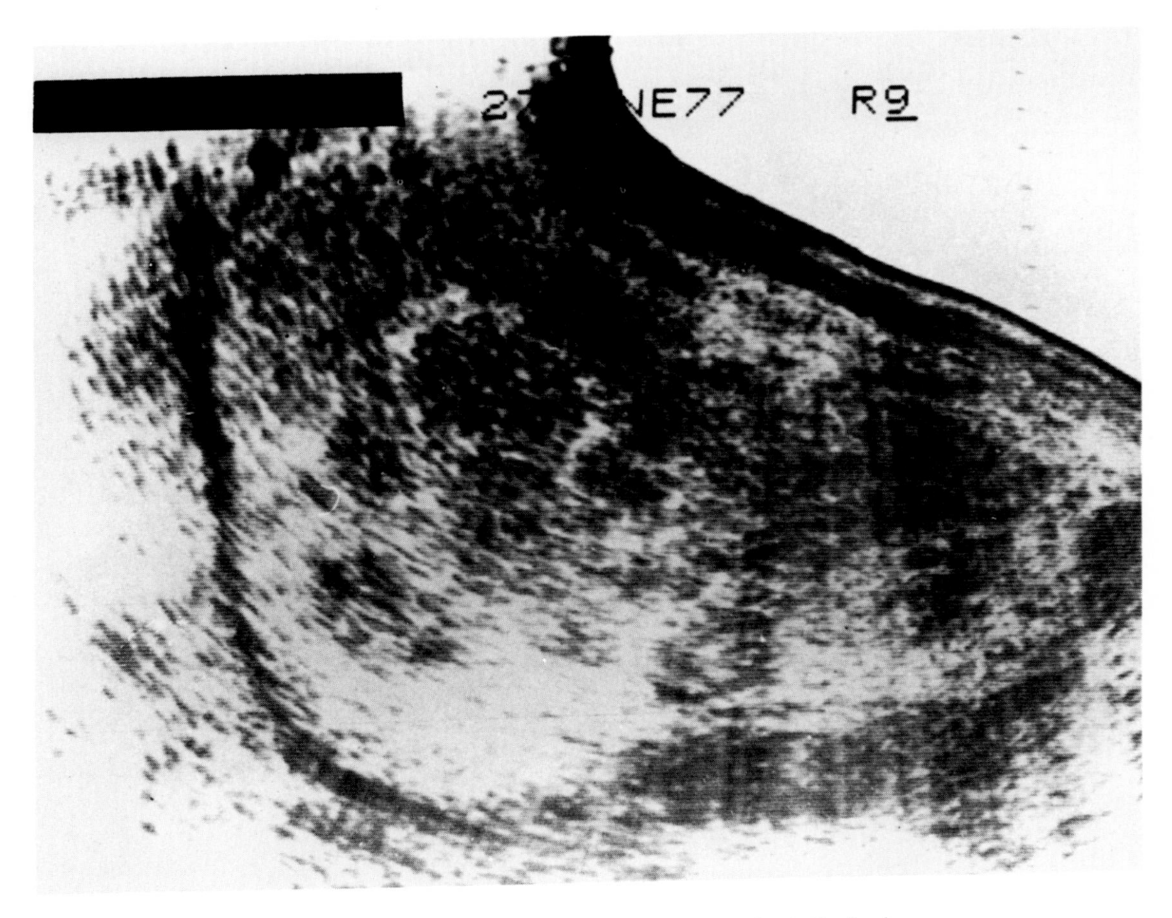

Figure 11. Multiple "bull's-eye"-type metastatic foci.

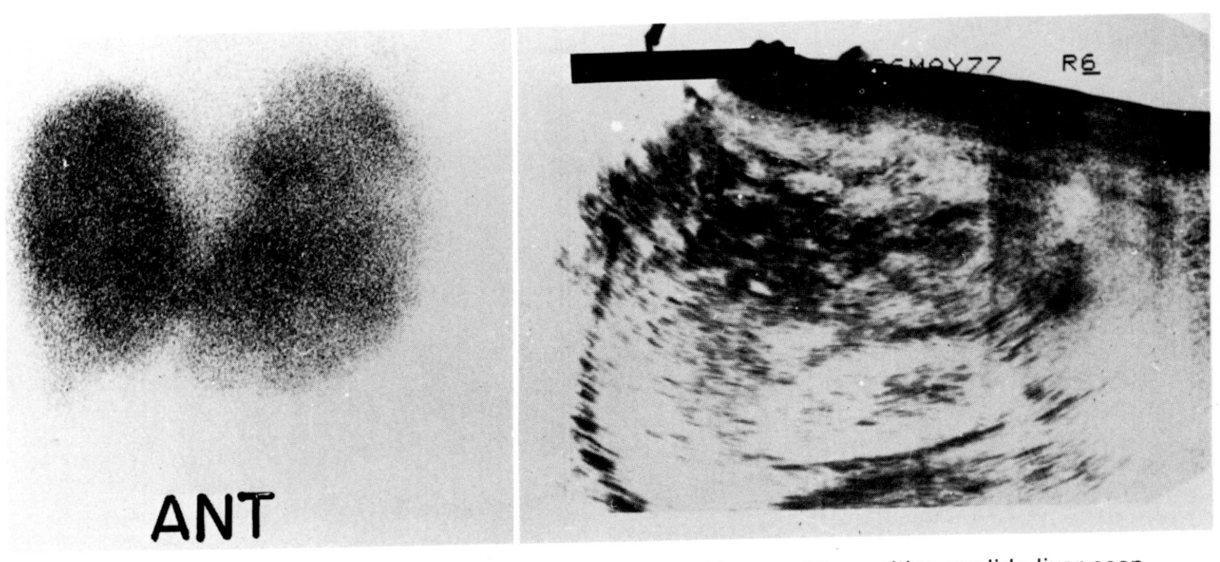

Figure 12. "Mixed echo pattern" of metastatic disease. Nonspecific positive nuclide liver scan.

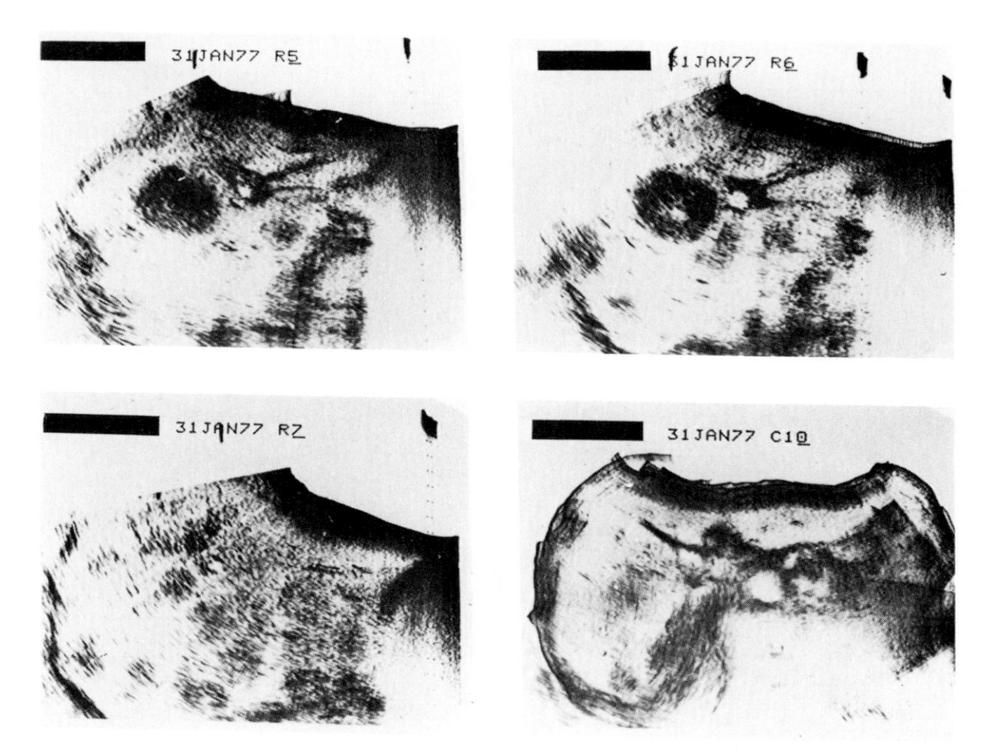

Figure 13. "Focal dense echo pattern" of metastatic disease. Frequently seen in GI tract adenocarcinomas.

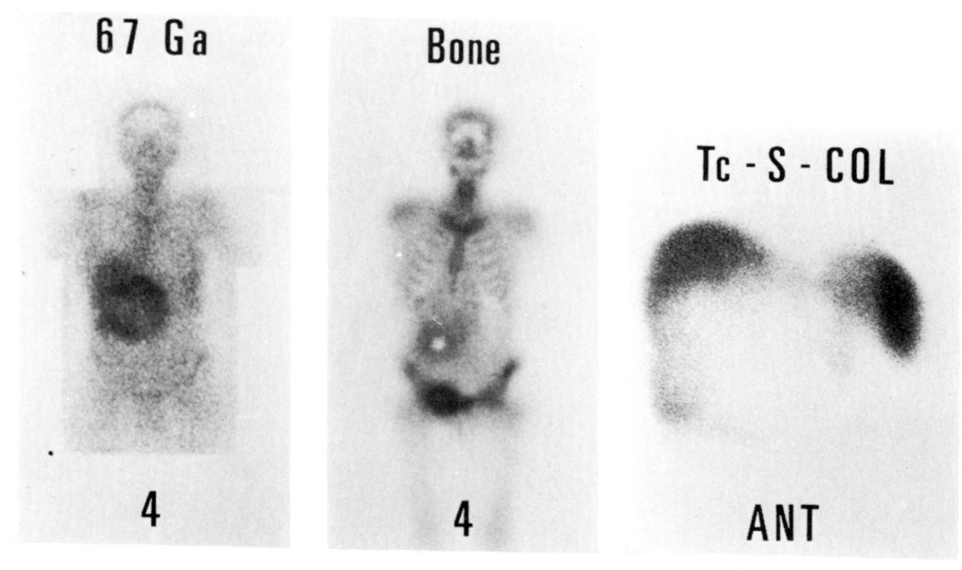

Figure 14. "Focal decreased echo pattern" of metastatic disease. Gallium and Tc-99m MDP positive mucin-secreting adenocarcinoma of colon (marker over lesions).

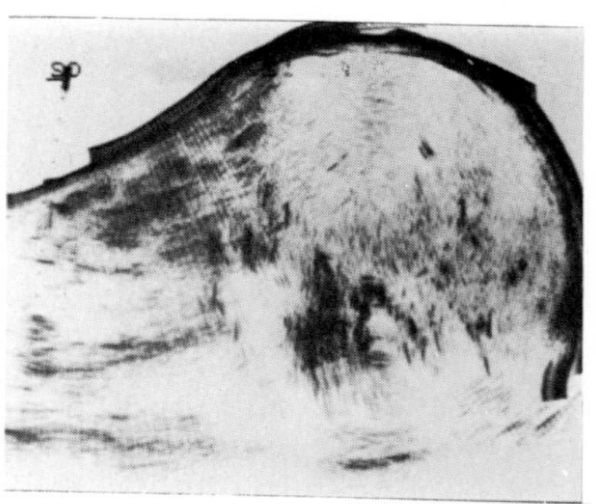

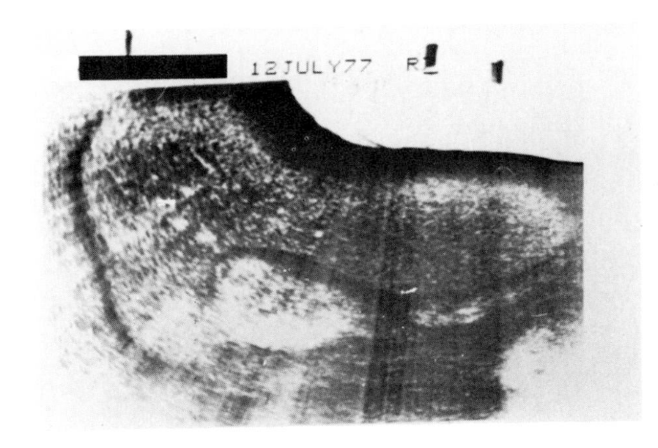

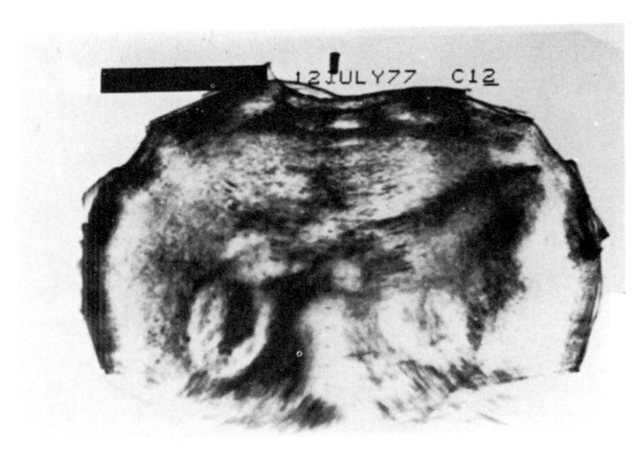

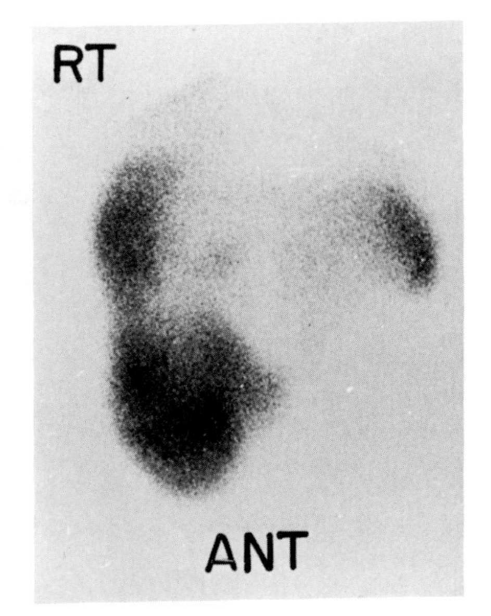

Figure 15. Positive liver nuclide scan and ultrasound demonstration of intraheptic *diffuse* dense echo pattern of cirrhosis.

pattern." This appearance has been frequently seen in patients with GI tract tumor metastases, but is still felt to be nonspecific.

The fourth pattern, and the least frequently seen, is the "focal decreased echo pattern" in the lower right-hand corner of Figure 14. Again, the technetium sulfur colloid liver image reveals a rather large filling defect, which we see in the upper left-hand corner, is gallium positive, and even shows uptake on a technetium methylene diphosphonate bone scan in the upper right-hand corner. With this much technical information, one can certainly suggest the proper diagnosis of mucin-secreting adenocarcinoma of the colon.

Figure 15 shows an anterior technetium sulfur colloid liver image with apparent destruction of the left lobe of the liver. Ultrasound examination in this case, however, reveals diffuse dense echoes throughout the liver, which is a pattern associated with Laennec's cirrhosis, which this patient has. Ultrasound in this case suggested the proper diagnosis, though without the information on the nuclide liver image, the ultrasound image would have been nonspecific.

Figure 16 shows a patient with an enlarged liver and multiple filling defects; again, nonspecific but suspicious for multiple metastases. Ultrasound in this case reveals multiple cystic lesions in a patient with polycystic disease of the liver and kidneys.

Figure 17 demonstrates the additional value and complimentary role of ultrasound in a patient with a markedly abnormal liver scan with multiple filling defects, including a large posterior filling defect. Notice that the large posterior filling defect is actually a large right renal cyst and the multiple small filling defects are actually dilated intrahepatic bile ducts. The patient was jaundiced and did have carcinoma of the colon, and it was felt that the liver defects on the isotope scan were most probably metastases. After the ultrasound findings, percutaneous cholangiography was done and a large common duct stone was found. This was removed surgically and the patient recovered without difficulty.

As is demonstrated, ultrasound imaging of the liver and spleen adds specificity to the sensitive nuclide liver scan.

Figure 18 is a suggested approach in thought process when one is confronted with an abnormal or equivocal liver scintigram. In this outline, use of gallium scanning is also suggested as this modality can separate the simple cyst from the liver abscess in practically all cases. We feel that the proper approach to the work-up of a potential liver abnormality is first the technetium sulfur colloid liver image to give the best global approach, and when lesions are found or when the nuclide liver image is indeterminant, ultrasound scans play a role in adding specificity to the sensitive scanning tool.

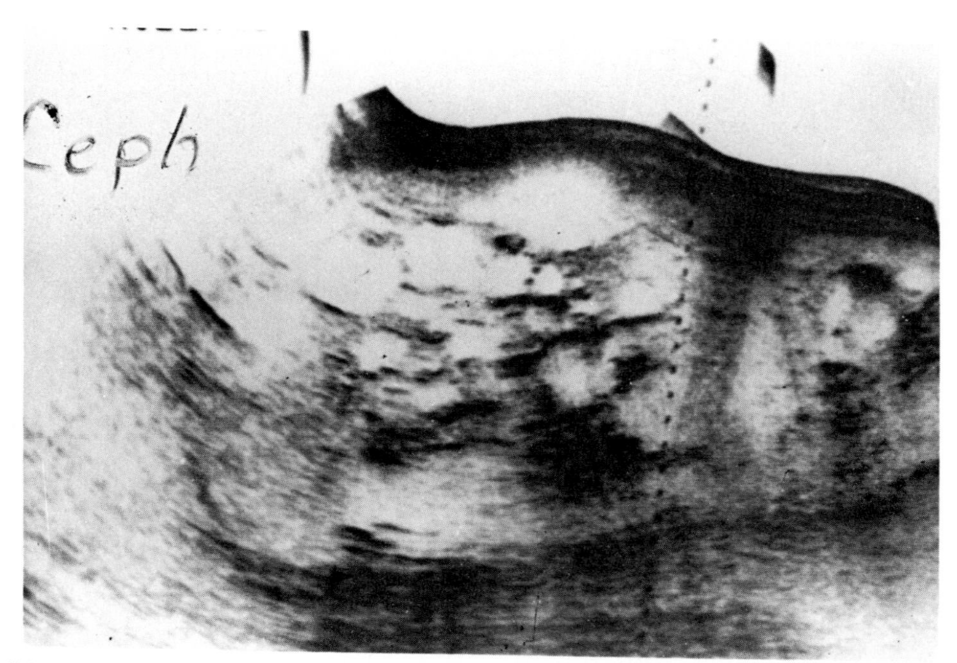

Figure 16. Nuclide liver scan was positive and suspicious for multiple metastases. Ultrasound showed multiple cysts in the liver polycystic disease.

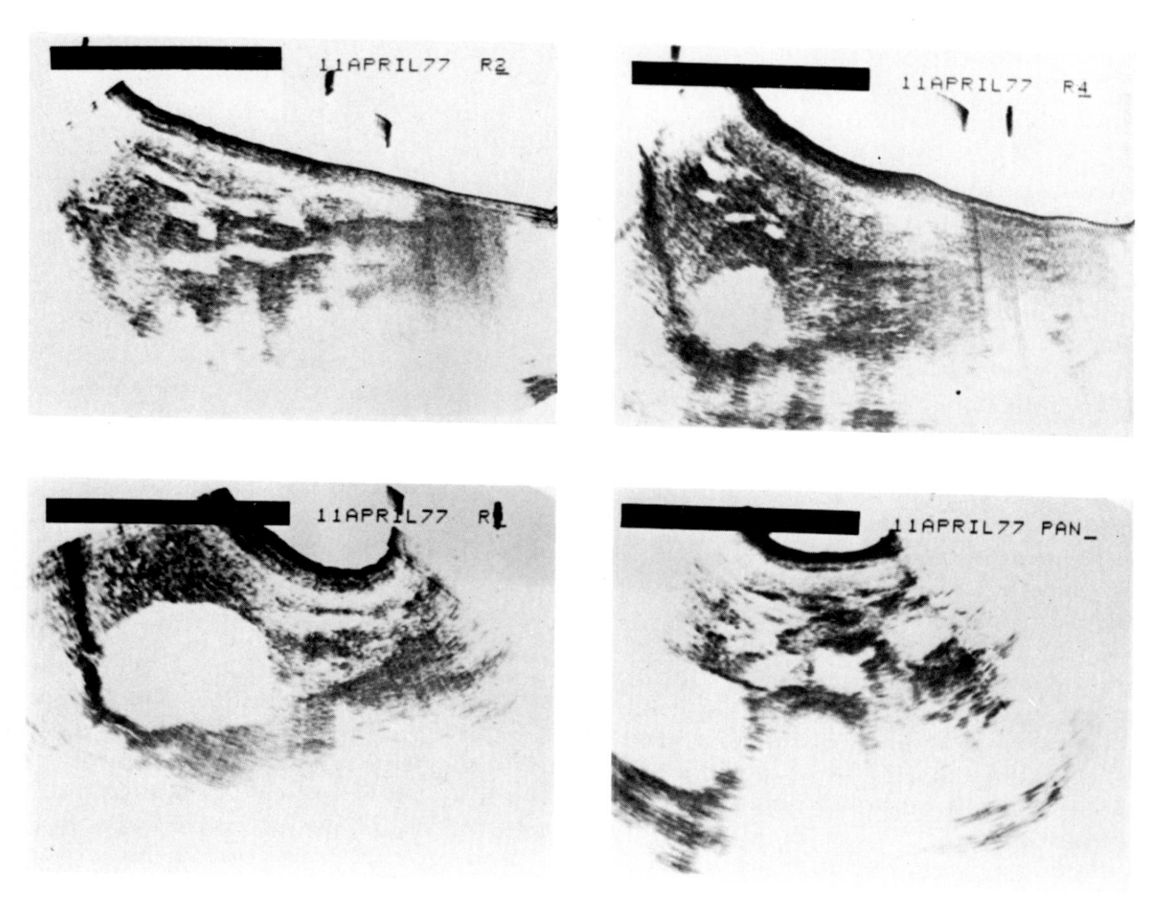

Figure 17. Nuclide liver scan positive in patient with carcinoma of the colon. Defect shown to be dilated bile ducts due to gallstone and large right renal cyst by ultrasound.

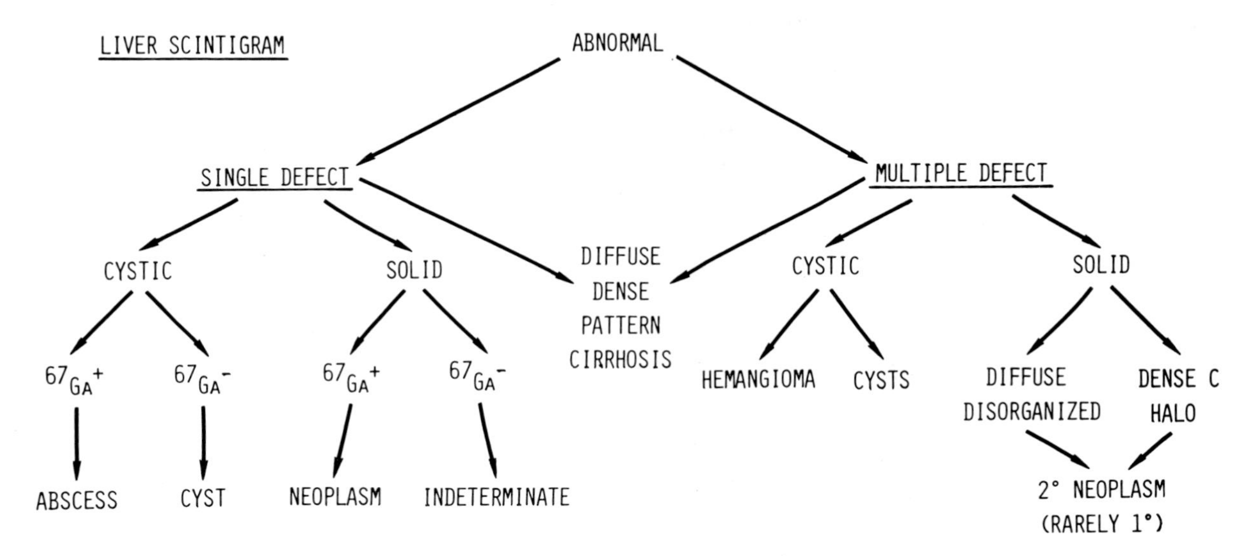

Figure 18. Ultrasound adds specificity.

Radionuclide Imaging, Computed Tomography, and Gray-Scale Ultrasonography of the Liver

Brian W. Wistow
Zachary D. Grossman

Analysis of the strengths and weaknesses of radionuclide imaging, computed tomography, and gray-scale ultrasonography for detection of space-occupying processes in or immediately adjacent to the liver suggests that biochemical liver-function tests and radionuclide imaging should probably remain the initial screening examinations. Opinions vary as to the necessity of a "second screen." If one does wish to further decrease the already low false negative rate of combined nuclear imaging and biochemical assays, then ultrasound is the procedure of choice, since contrast material is not necessary, expense is lower than CT, and ionizing radiation is not involved. Moreover, ultrasound is effective for clarification of equivocal or positive nuclear images.

CT, at present, serves best to characterize lesions detected by the other modalities, or to clarify equivocal nuclear images when ultrasound is not feasible due to technical factors (body habitus, intestinal gas, external impediments, etc.). The superior resolution of fast (18 to 20 seconds) CT units compared to both radionuclide imaging and ultrasound is well documented, but the clinical value of reducing the already low false negative rate of combined radionuclide imaging, biochemical assay, and ultrasound is doubtful. Further comparative studies, particularly when even higher resolution and faster scanners become available, will be necessary.

Both CT and ultrasound are superior to nuclear imaging as the initial screening procedure for suspected noninflammatory biliary tract obstruction. The new Tc-99m IDA derivatives may assume a prominent role in effectively excluding acute cholecystitis by demonstrating cystic duct patency.

Space-Occupying Lesions

Noninvasive detection of space-occupying processes in or immediately adjacent to the liver is feasible by three imaging modalities—radionuclide scintigraphy, ultrasound, and computed tomography. In addition, biochemical tests may raise the suspicion of intrahepatic lesions. To best utilize limited health-care funds and facilities, the relative roles of the three imaging modalities and the value of images relative to biochemical tests must be explored. In addition, a larger question—whether detection of space-occupying processes within the liver by any means (imaging or biochemical) is of prognostic importance—requires attention.

While standard nongated gamma cameras, "slow" CT units with 2 to 3.5-minute scanning times, and gray-scale sonography have been systematically compared,[1,2] no organized comparative data using "fast" CT (18 to 20 seconds scanning time) are available. Therefore, approaches to the detection and characterization of space-occupying processes with the lastest equipment derive from piecemeal, almost anecdotal information.

Advantages and Disadvantages of Each Modality

Nuclear imaging with "state-of-the art" gamma cameras, high resolution collimators, adequate doses of technetium-99m sulfur colloid, and elimination of respiratory motion suffers from the poorest spatial resolution. Because normal liver tissue accumulates Tc-99m sulfur colloid, deep lesions—surrounded by radioactive normal tissue on all sides—are more difficult to detect than superficial ones. On the other hand, the imaging procedure is routine, with little dependence upon operator skill, and the gamma camera views all parts of the liver, without interference from excessive obesity, intestinal gas, or overlying ribs.

The spatial resolution of ultrasound is clearly superior to that of radionuclide imaging in areas of the liver accessible to the transducer (Fig. 1), so that a negative

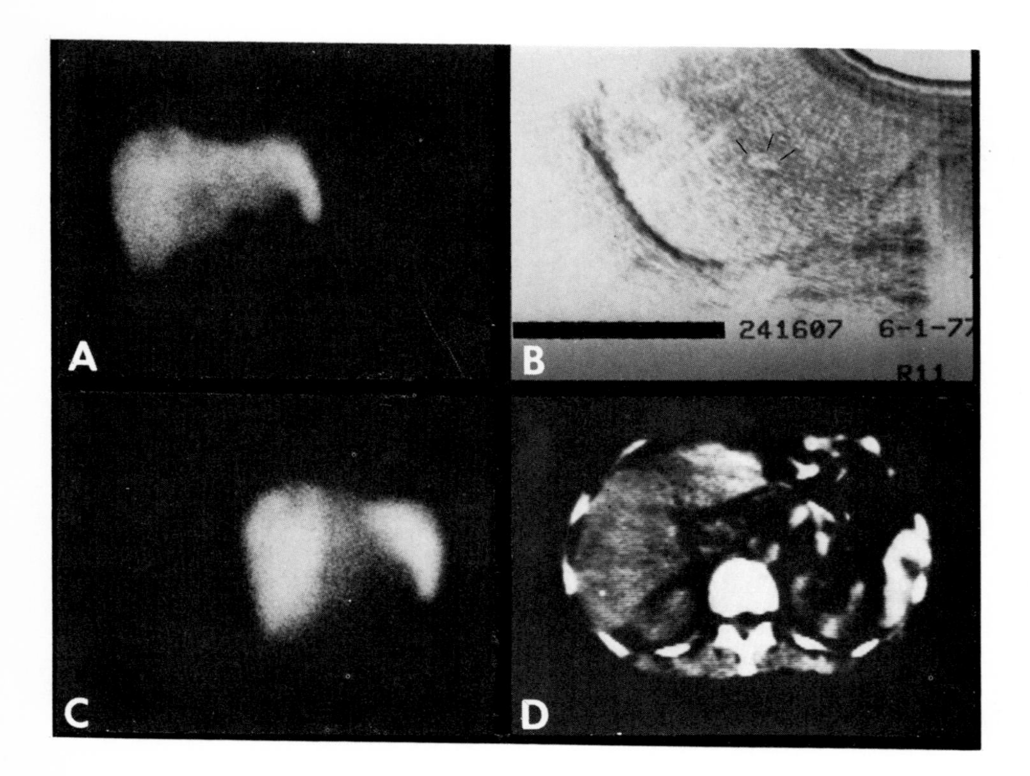

Figure 1. Metastases: **A,C.** Anterior and posterior radionuclide images reveal a prominent space-occupying process in the superior portion of the right lobe, adjacent to the diaphragm, as well as stongly suspicious decreased colloid accumulation in the periportal region. **B.** The longitudinal sonogram, 11 cm to the right of the midline, demonstrates the larger superior lesion adjacent to the diaphragm. In addition, a tiny deep lesion (markers) with a necrotic center is present. **D.** Computed tomography establishes that, in addition to the lesions demonstrated by nuclear imaging and sonography, multiple well-resolved tiny foci are present throughout, best appreciated in the lateral aspect of the right lobe.

radionuclide study may occasionally be followed by a positive ultrasound.[3–9] In addition, the porta hepatis, gallbladder bed, and left lobe are particularly well demonstrated by sonography. Since many equivocal scintigrams result from the gallbladder fossa, the porta hepatis, or extrinsic compression of the left hepatic lobe, ultrasound is frequently able to definitely clarify the nuclear findings (Fig. 2). Production of diagnostic-quality images, however, is highly dependent upon operator skill, and certain hepatic regions—e.g., the "dome" of the right lobe under the rib cage—may be inaccessible, even to the expert (Fig. 3). Moreover, excessive obesity, bowel gas, patient motion, surgical wounds, and surgical dressings inhibit ultrasonography.

Like radionuclide imaging, CT examines the entire organ in virtually all patients, without regard to external impediments—drains, obesity, etc.—and, like ultrasound, the spatial resolution of CT is superior to that of gamma scintigraphy. Recent studies utilizing the 18-second CT scanner suggest that the spatial resolution of CT may be superior to sonography even in areas accessible to ultrasound and examined by a skilled op-

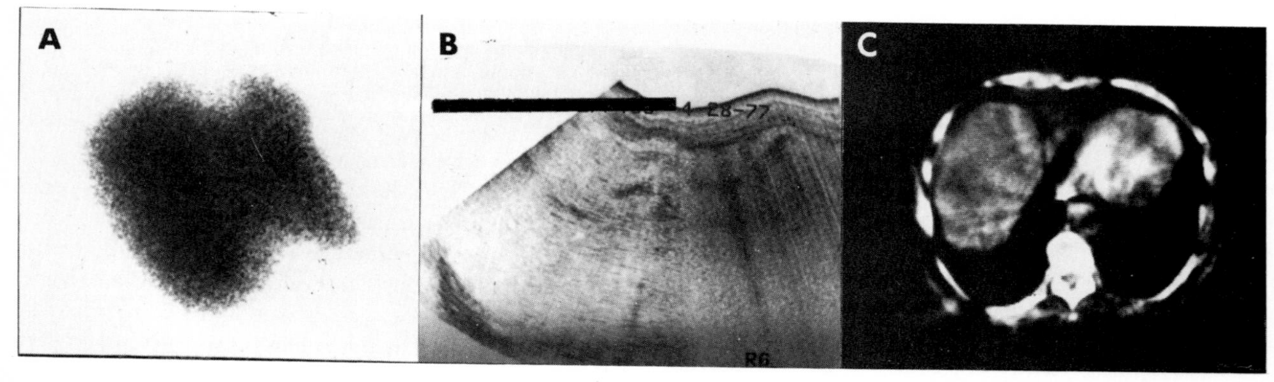

Figure 2. Metastasis: **A.** The lateral radionuclide image reveals an unequivocal lesion adjacent to the diaphragm, in the "dome" of the right hepatic lobe. **B.** Although the lesion in **A** is sufficiently large to be easily resolved by sonography, anatomic considerations prevent sonographic visualization. **C.** CT with the 18-second scanner clearly demonstrates the right-lobe lesion in **A.** The slower CT units were often ineffective in this location, due to diaphragmatic motion. Multiple smaller defects in the right lobe in this case may not be reliably considered lesions, since averaging of liver and lung attenuation coefficients at this anatomic level may produce artifacts of the identical pattern.

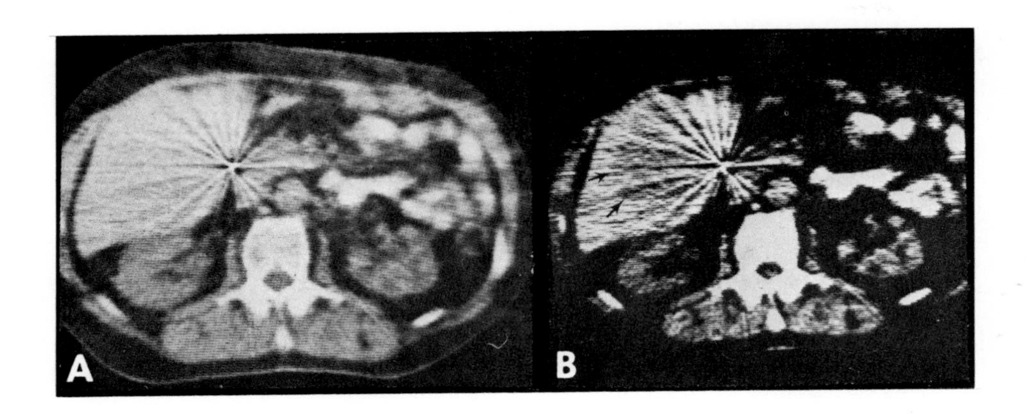

Figure 3. A. The gray "wide-window" (600) image reveals no intrahepatic lesions. Note the stellate pattern of surgical-clip artifact. **B.** The identical image viewed with a "narrow window" (180) strongly suggests the presence of a prominent, solitary space-occupying process (arrows). Nearly isodense lesions—such as this hepatoma—are most often overlooked when the images are analyzed with wide-window settings.

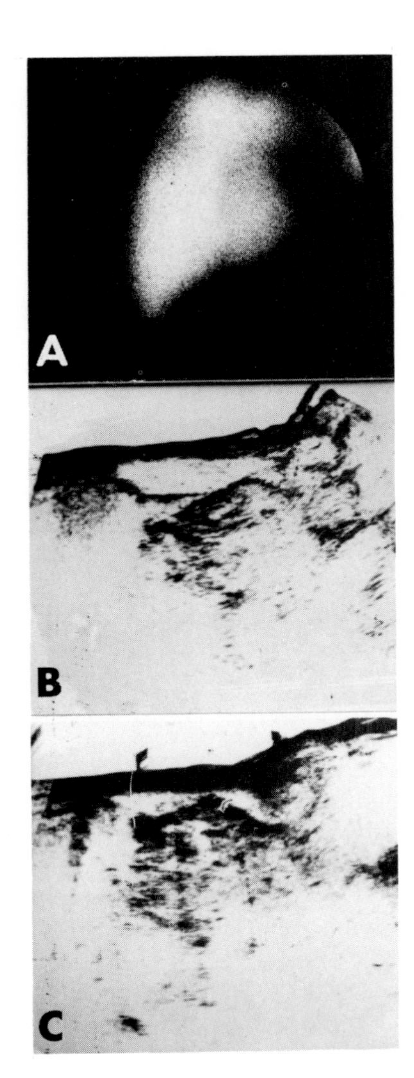

Figure 4. A. The anterior radionuclide image reveals an apparent impression upon the left lobe, concave medially. **B,C.** Longitudinal sonograms (head to viewer's right) establish that the left lobe thins markedly at the midline. No intrinsic or extrinsic lesions are present, and the thinning represents a normal variant.

erator (Fig. 1). Unlike the other modalities, however, a full-scale CT examination is "invasive," because an intravenous injection of contrast medium is necessary for maximum efficacy. Reports of small lesions missed by unenhanced CT are common. Motion artifact, surgical clip artifacts, and the use of inappropriately "gray" images rather than more appropriate narrow-window images (Fig. 4) are also detrimental, but these difficulties are rapidly fading as equipment advances and interpreter skills increase.[10]

The Incidence of False Negative and False Positive Studies

To date, no study has established the true false positive and false negative rates of radionuclide imaging, because biopsies rarely follow a liver scan reported as "negative".[11,12] Postmortem examinations almost invariably have followed weeks, if not months, after a negative liver scan, and thus the true status of the liver at the time of the original imaging procedure has remained unknown. Moreover, while laparotomy for resection of a known intraabdominal neoplasm has frequently followed a negative liver study, inspection of the liver during surgery—without actual biopsy—has been the usual procedure; thus, the presence or absence of small, deep lesions not detected by the radionuclide image has not been established. CT and gray-scale ultrasound, however, provide independent methods of assessing the liver considered "negative" by the radionuclide method. In two previous studies, sonographic and CT examinations considered "unequivocally positive" by a panel of six observers were accepted as representative of space-occupying processes, even without histologic proof.[1,2] Other studies of hepatic ultrasound and computed tomography have generally required biopsy proof in almost all cases; however, none of these studies have involved all three imaging modalities, and most have been limited to two or even one method. To rigorously determine false negative rates of radionuclide images, livers must be examined, within approximately 1 week, by all modalities, and hepatic tissue—either through transcutaneous biopsy or laparotomy—must be obtained. Such a study has not appeared to date.

The Relative Importance of False Positive and False Negative Examinations

Although false negatives have traditionally haunted radiologists—in the form of missed fractures, missed colonic polyps, etc.—the false positive radionuclide liver examination may, in fact, represent a greater danger than the false negative study. False positive examinations may inhibit the surgeon from resecting a known primary lesion, and instead of curative intervention only palliative intervention (or no intervention) may result. Thus, a false positive study is associated with the risk of losing some potentially salvageable patients. Fortunately, both ultrasound and CT are effective in clarifying the normal variants and extrinsic impressions which produce most false positive or equivocal radionuclide images. False negative radionuclide studies, on the other hand, result in a curative surgical attack on a primary lesion which has, in fact, already metastasized. Although the expense and morbidity of such surgery is clearly undesirable, the potential loss of life is much smaller than that resulting from false positive radionuclide images. Furthermore, many such resections would be attempted in the abdomen, and the liver would be inspected by the surgeon prior to resection of the primary lesion. Such inspection would reveal some hepatic metastases missed on the false negative radionuclide examination, since tiny deep lesions may also be associated with small surface lesions. Accordingly, plans for curative resection of an abdominal primary tumor following a false negative liver scan would often be aborted at the time of surgery, and the procedure would be limited to palliation. When liver function tests as well as the radionuclide scan are negative, the number of false negatives missed by both the biochemical assay and radionuclide imaging combined may be as low as 10 percent, and since many of the primary lesions in this 10 percent would be located within the abdomen, few unnecessary "curative" procedures would actually be attempted.

The Relationship of Imaging Procedures to Biochemical Tests

In addition to the lack of information regarding false negative radionuclide hepatic images, widely varying opinions abound regarding the value of standard liver function tests (alkaline phosphatase, SGOT, bilirubin, albumin, globulin) for the detection of metastatic disease.[12–14] A large review of radionuclide images produced by rectilinear scanning failed to reveal a single case of normal liver function tests in the presence of even one definitively demonstrated space-occupying metastatic carcinoma.[15] With the advent of higher resolution gamma cameras, however, and with the detection of smaller lesions, many clinically oriented nuclear physicians believe that such cases—i.e., normal liver function tests in the presence of proven space-occupying metastatic liver disease—are common, if not routine. Surprisingly, however, review of the literature reveals few documented cases of this nature. Moreover, not a single published incidence of normal liver function tests combined with a normal radionuclide scan, in the presence of proven small metastatic deposits, is available. Again, despite lack of documentation, many practicing nuclear physicians and some surgeons and gastroenterologists believe that such a combination would not be uncommon. Thus, a lack of information regarding true false negative rate of radionuclide examinations and the efficacy of liver function tests for the detection of space-occupying metastatic processes within the liver inhibits the development of a coherent, generally agreed-upon approach to the liver "workup."

Conclusion

Overall, then, a definitive plan for hepatic "workup" by medical imaging cannot be formulated with certainty because of insufficient data regarding the false positive and false negative rates of the three modalities. Moreover, rapid technologic advances in either biochemical methodology or imaging might rapidly and drastically alter any formulated plan. However, for the present:

1. Radionuclide imaging and biochemical tests should probably remain the initial screening examinations for suspected nonobstructive intrahepatic lesions.
2. If the radionuclide scan is normal, and if routine liver function tests are unremarkable, then ultrasound as a "second screen" may be performed, if one wishes to further reduce the already low false negative rate of combined nuclear imaging and biochemical tests.
3. When the screening nuclear images are positive or equivocal, ultrasound has proved effective for clarification (Fig. 5). Since contrast material is not necessary, expense is lower than CT, and ionizing radiation is not involved, ultrasound remains the procedure of choice to follow the screening radionuclide study.
4. CT, at present, serves best to characterize hepatic lesions detected by the other modalities, or to clarify

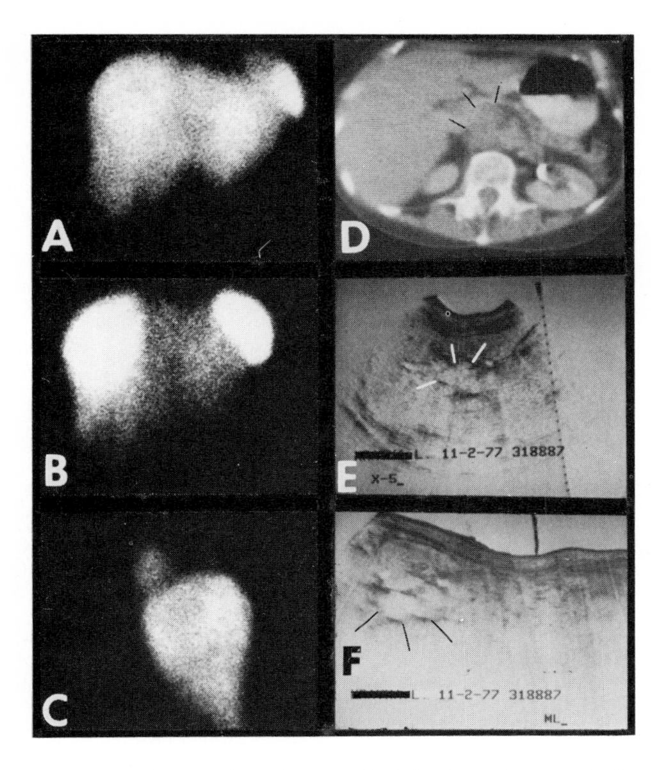

Figure 5. Proven lymphoma: **A,B,C.** Anterior, posterior, and right lateral radionuclide images reveal prominence of the porta hepatis and decreased radiocolloid accumulation—probably secondary to extrinsic compression—in the left lobe. **D.** Computed tomography reveals periaortic adenopathy, obliterating individual planes between the retroperitoneal vessels. **E,F.** Transverse and longitudinal sonograms confirm the presence of gross adenopathy revealed in **D.** Sonography and CT are equivalent in this case.

equivocal nuclear studies when performance of sonography is not feasible.

5. The justification of enhanced CT (when the radionuclide study, liver function tests, and ultrasound are normal) is doubtful. Further comparative studies with histologic proof and long-term patient followup will be necessary for clarification.

The Hepatobiliary System

Since the oral cholecystogram is noninvasive and effective, it remains the primary tool for detection of calculi. When the gallbladder fails to visualize, however, and when gallbladder calculi are clinically probable, then ultrasound may be extremely useful for their detection.[16] Although the newer hepatobiliary radiopharmaceuticals in the IDA-derivative group frequently demonstrate the gallbladder with clarity, limited resolution prevents the detection of calculi. Because the section thickness of individual CT images is much greater than that of individual tiny calculi, CT does not strongly compete with sonography as a "second line" detector of calculi following the "non-vis" oral cholecystogram.

In the diagnosis of gallbladder inflammation, as opposed to calculi, the radionuclide method may find wide application. Surgeons and pathologists agree that, in general, the possibility of an acute gallbladder inflammation in the presence of a patent cystic duct is remote. Therefore a noninvasive technique which provides gallbladder visualization within 20 minutes is highly desirable.[17,18] The oral cholecystogram requires a longer time period, and the contrast material might not be tolerated by patients already in gastrointestinal distress. Intravenous cholangiography, of course, is associated with significant morbidity and mortality. Since failure to visualize the normal gallbladder on a radionuclide cholecystogram in the fasting patient is rare, gallbladder nonvisualization virtually establishes the presence of cystic duct obstruction. Conversely, definite gallbladder filling establishes duct patency, excluding the diagnosis of acute cholecystitis (Fig. 6). The use of the radionuclide method of exclusion of acute cholecystitis has not found wide acceptance to date, presumably because of lack of pharmaceutical availability and lack of information in the medical community regarding its use. Sonographic localization of the gallbladder and palpation to establish the presence or absence of point ten-

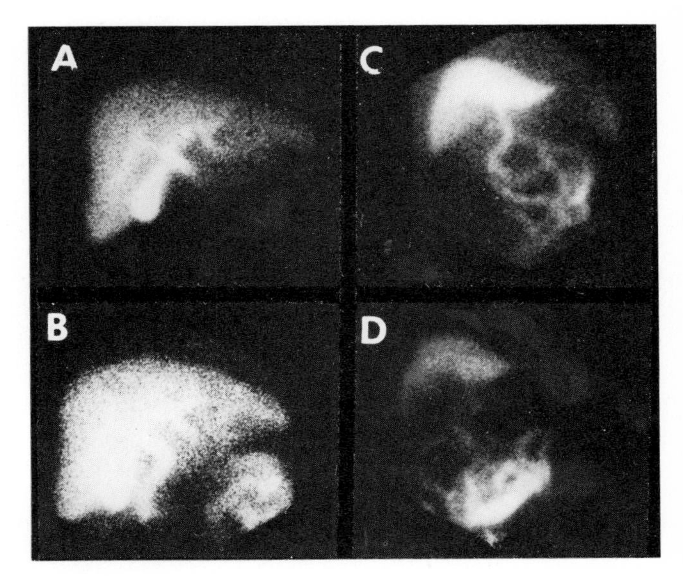

Figure 6. Normal: **A,B.** Here 20- and 50-minute hepatobiliary images using a Tc-99m-labeled IDA derivative demonstrate the gallbladder and the jejunum, establishing the presence of cystic and common duct patency. Acute cholecystitis: **C,D.** Comparable images of a patient with acute cholecystitis, utilizing the identical radiopharmaceutical, reveal patency of the common duct. However, the gallbladder is not observed, strongly supporting the likelihood of cystic duct obstruction.

derness is also a promising technique but requires a highly-skilled physician operator and may not be feasible when the gallbladder is deep or retrohepatic.

In the diagnosis of duct dilatation, both ultrasound and CT offer significant advantages over any other techniques.[19-23] Percutaneous transhepatic cholangiography is clearly unacceptable as a first procedure, as is endoscopic retrograde cholangiography. The intravenous cholangiogram is only occasionally effective. While the radionuclide examination may confirm the presence of dilated ducts suspected on the standard scintigram, the examination is unlikely to produce the striking detail available with both ultrasound or CT. In addition, the ability of CT and sonography to reveal ductal ectasia depends only upon ductal size, whereas the intravenous radionuclide cholangiogram requires continued hepatic function. At elevated bilirubin levels, the "anatomic" modalities are superior to the "functional" one. Therefore, either ultrasound or CT would serve best as the initial examination in patients with suspected biliary tract obstruction. If the nature and site of the obstruction are not established by ultrasound first and CT subsequently, then percutaneous transhepatic cholangiography would logically follow. When the transcutaneous procedure is contraindicated, endoscopic retrograde cholangiopancreatography might be of value. CT and ultrasound, then, provide excellent diagnostic visualization of the biliary tree, and only if these procedures fail to reveal the etiology of ectasia do other, more invasive procedures become necessary.

In summary, the relative roles of CT, ultrasound, and the Tc-99m IDA derivatives for study of the hepatobiliary tract have been well outlined by Ronai[24] to wit:

1. CT and ultrasound remain superior as the initial screening procedures for suspected biliary tract obstruction.
2. Following CT or ultrasound, the percutaneous transhepatic cholangiogram may be indicated to show the site of obstruction.
3. When contraindicated, the percutaneous study may be replaced by endoscopic retrograde cholangiopancreatography.
4. The oral cholecystogram is presently the screening procedure of choice for gallbladder calculi.
5. In cases of gallbladder nonvisualization following oral cholecystography, ultrasound effectively demonstrates many calculi.
6. The Tc-99m IDA derivatives may effectively exclude acute cholecystitis in fasting patients by demonstrating cystic duct patency.
7. A skilled physician-sonographer may establish the presence or absence of point tenderness over the gallbladder, providing strong evidence for or against acute cholecystitis.

References

1. Grossman ZD, Wistow BW, Bryan PJ, et al: Radionuclide Imaging, Computed Tomography and Gray Scale Ultrasonography of the Liver. A Comparative Study, J Nucl Med 18:327, 1977
2. Bryan PJ, Dinn WM, Grossman ZD, Wistow BW, McAfee JG, Kieffer SA: Correlation of Computed Tomography, Gray Scale Ultrasonography and Radionuclide Imaging of the Liver in Detecting Space-Occupying Processes, Radiology 124:387, 1977
3. Sample WF, Po, JB, Poe, ND, et al: Correlative Studies between Multiphase Tomographic Nuclear Imaging and Gray Scale Ultrasound in Extra- and Intrahepatic Abnormalities. In White DN (ed.): Ultrasound in Medicine. New York, Plenum, Vol. 2, 1976, pp. 175-176
4. Sample WF, Editorial: Non-invasion Hepatic Imaging, Appl Radiol 6:123, 1977
5. Taylor KJW, Carpenter DA, Hill CR, et al.: Gray Scale Ultrasound Imaging. The Anatomy and Pathology of the Liver, Radiology 119:415, 1976
6. Taylor KJW, Sullivan D, Rosenfeld AT, Gottschalk A: Gray Scale Ultrasound and Isotope Scanning: Complementary Techniques for Imaging the Liver, Am J Roent 128:277, 1977
7. Sanders DA, Sanders RC: The Complementary Use of B-Scan Ultrasound and Radionuclide Imaging Techniques, J Nucl Med 18:205, 1977
8. Taylor KJW, Rosenfeld AT: Nuclear Medicine vs. Ultrasound, J Nucl Med 18:1138, 1977
9. Sanders DA, Sanders RC: Reply to "Nuclear Medicine vs. Ultrasound," J Nucl Med 18:1139, 1977
10. Stephen DH, Sheedy PF, Hattery RR, MacCarty RL: Computed Tomography of the Liver, Am J Roent 128:579, 1977
11. Ludbrook J, Slavotinek AH, Ronai PM: Observer Error in Reporting on Liver Scans for Space-Occupying Lesions, Gastroenterology 62:1013, 1972
12. Rosenthal SN: Are Hepatic Scans Overused? Digest Dis 21:659, 1976
13. Sisson JC, Schoomaker EB, Ross JC: Clinical Decision Analysis. The Hazard of Using Additional Data, JAMA 236:1259, 1976
14. Rosenthal S, Kaufman S: The Liver Scan in Metastatic Disease, Arch Surg 106:656, 1973
15. McAfee JG, Ause RG, Wagner HN Jr: Diagnostic Value of Scintillation Scanning of the Liver, Arch Int Med 116:95, 1965
16. Leopold GR, Amberg J, Gosink BB, Mittelstaedt C: Gray Scale Ultrasonic Cholecystography: A Comparison with Conventional Radiographic Techniques, Radiology 121:445, 1976
17. Ryan J, Cooper M, Loberg M, Harvey E, Sikorski S: Technetium-99m-labeled *N*-(2,6 -Dimethylphenylcarbamoylmethyl) Iminodiacetic Acid (Tc-99m HIDA): A New Radiopharmaceutical for Hepatobiliary Imaging Studies, J Nucl Med 18:997, 1977
18. Wistow BW, Subramanian G, Van Heertum RL, et al: An Evaluation of ^{99m}Tc-Labeled Hepatobiliary Agent, J Nucl Med 18:455, 1977
19. Taylor KJW, Carpenter DA, McCready VR: Ultrasound

and Scintigraphy in the Differential Diagnosis of Obstructive Jaundice, J Clin Ultrasound 2:105, 1974
20. Taylor KJW, Rosenfeld AT: Gray-Scale Ultrasonography in the Differential Diagnosis of Jaundice, Arch Surg 112:820, 1977
21. Milini S, Sable J: Ultrasonography in Obstructive Jaundice, Radiology 123:429, 1977
22. Stanley RJ, Sagel SS, Levitt RG: Computed Tomography of the Body: Early Trends in Application and Accuracy of the Method, Am J Roent 127:53, 1976
23. Stephens DH, Hattery RR, Sheedy PF: Computed Tomography of the Abdomen: Early Experience with the EMI Body Scanner, Radiology 119:331, 1976
24. Ronai PM: Hepatobiliary Radiopharmaceuticals: Defining Their Clinical Role Will be a Galling Experience. Editorial, J Nucl Med 18:488, 1977

Discussion: Liver Imaging

Moderator: Atis K. Freimanis
Panelists: Edward A. Lyons
Wayne W. Wenzel
Brian W. Wistow

FREIMANIS: In making a point, Dr. Lyons mentioned reference marks—painting lines on the skin or otherwise identifying section levels. I have found that some reference marks are very helpful; indeed, it is quite important to know how the organs are related to each other. However, in consulting on CT scans referred to us from outside institutions, it is difficult to correlate the scans if the marking systems are different for the various modalities. Therefore, we are proposing that there be a uniform system of level or section markings on all of these examinations. Any comment from the panel on that? Do you use any particular systems in reference to body surface landmarks?

LYONS: There is no question that body surface landmarks are important and are used all the time. The point I was trying to make is that you should not rely upon them exclusively in trying to identify specific organs because of the tremendous variability. The surface landmarks that we use are: the midline, the iliac crest, the symphysis pubis, and the xyphoid. But it is difficult to relate those landmarks from patient to patient and sometimes from study to study in one patient.

FREIMANIS: How large were the hemangiomas you echoed and labeled cystic? We had echoed three giant, that is, over 15 cm liver hemangiomas and all were solid. A single small hemangioma was cystic. Would you have a comment on hemangioma?

WENZEL: Yes. First of all, I agree that they could look solid. We have only had one in my entire practice which was 2 to 3 cm in size and appeared cystic, but others that we have seen from other institutions also have appeared cystic on ultrasound. But I think that with the improvement of gray scale, these will probably show the pattern that goes with them. So, realistically, though they are very vascular structures, they do have some solid components and probably will show some low-grade echoes within them.

FREIMANIS: I have a written comment here concerning hemangiomas. Flow studies may be falsely negative in liver hemangiomas. These lesions, however, will fill on contrast angiography. That is, of course, a subject we have not discussed much today. There is still a major place for angiography. In our experience, if applied specifically, it has been most helpful in clarifying abdominal lesions of various kinds.

QUESTION: Please describe your method of obtaining a nuclear liver scan only at times of stopped respiration, in other words, respiratory gating.

WISTOW: Several different systems are available commercially. I am not familiar with very many of them. There are two basic different types of respiratory gating systems. One uses the patient's own respiration by putting two transducers on either side of the chest; the change in impedance, or whatever it would be, tells the machine the various phases of respiration. Just pushing a button that says "80 percent" makes the machine accept counts only during the time when the curve is closest to expiration or to inspiration. One can pick any point on the curve. This is a very basic system and it is very slow since counts are taken only during maybe one-fifth of the time that the camera could be collecting them. We do not use it very often and I really mention it for interest. We have not really found it very useful. Electronic gating systems are becoming available. These systems will calculate perhaps the center of activity or the *apparent* center of activity of the liver and as the liver moves up and down will electronically shift the position of the dots. But I think the liver is so variable and moves in so many different positions that this, I do not believe, will be particularly useful either. Does anybody else have comments?

WENZEL: The one you described last is the one we use now. It works. But so far, it has not helped me find a lesion. I find that dismaying because other people tell me that in their practices (especially Dr. Fordham yesterday—and I really believe what he says) there is unequivocal difference. Dr. Alex Gottschalk, from whom I happened to buy this instrument, tells me that it is most likely quite good. But so far, I can barely tell the difference. The images appear a little more crisp; we do everything we can, including taking the images on small field-of-view camera, high resolution, 750,000 counts, and everything else possible. I am more impressed with using the triple lens Polaroid system as being able to take a better look at the liver than I am with stop motion.

LYONS: I would just like to make one additional point. That is, during respiration, not only does the liver change position but it also may change shape. It is a very pliable structure.

I would further like to make one comment about hemangiomas and ultrasound. We have seen a number of small hemangiomas which appeared cystic even with recent equipment. But the larger ones appeared solid, very definitely solid.

With the newer equipment, I think that we really have to evaluate the way we look at the liver and the amount of information that is available from liver parenchyma. If one looks very closely at the texture of liver parenchyma on the ultrasound scan, one can identify metastatic lesions in the order of half a centimeter. Again, you have to be very, very careful. You have to have good equipment. And you have to pay a great deal of attention to very, very small changes in levels of gray. It is amazing to see the amount of metastatic disease showing just very, very small changes in these levels of gray.

So again, at this particular point in time, I think that even if the radionuclide and the CT scan are negative, it may still be possible to pick up diffuse metastatic disease on the ultrasound, although I do not have proof.

FREIMANIS: The next question is for Dr. Lyons. When you describe anatomical relationships to the inferior vena cava, are you using a perpendicular transducer position or do you angle cephalad? I suppose the question mostly would apply to transverse planes which might be angled cranially or caudally out, with the transducer being angled upward and then sweeping down, or vice versa.

LYONS: Maybe we can have some clarification on the question. There are about eight different ways you could answer it.

Are the kidney and the adrenal gland directly posterior to the inferior vena cava? Do you scan at right angles through the abdominal wall or how do you angle your transducer? That becomes important when you are talking about what is behind and what is in front.

Okay. Going from the anatomical sections, when I am talking about the structures behind the inferior vena cava, certainly—in a strictly perpendicular plane—the right adrenal lies directly behind the inferior vena cava and often in a child a normal adrenal is fairly large and will indent the inferior vena cava. The renal artery, of course, lies directly behind the IVC. The upper pole of the right kidney may or may not, in a normal condition, lie directly behind it, but a mass in the upper pole of the right kidney may elevate the posterior part of the right kidney and be then posterior to part of the inferior vena cava. The point that I really wanted to make is that distortion of the posterior part of the inferior vena cava may be caused by these three structures: the adrenal gland, the upper pole of the right kidney, and the right renal artery.

FREIMANIS: One should keep in mind that while the anatomical studies are extremely valuable and one needs to pursue them, the cadavers are often in a somewhat different state than the live person. Not only in terms of tissue texture but also in the location, relationships, and shape of the organs, the level of the diaphragm, and other normally movable structures. Thus, the intraabdominal organ relationships may be somewhat different in the cadaver. There is also considerable variability from one individual to another.

I have a statement here for discussion by Dr. Wistow. The writer disagrees with liver-lung nuclear imaging as the convention for diagnosis of subphrenic abscess. Do you know of a case of subphrenic abscess missed by gallium-67 scan?

WISTOW: I know only from the literature; in the presence of severe leukemia (no white blood cells to take up the gallium?). Subphrenic abscess should always be studied by either liver-lung scanning, computed tomography, ultrasound, or gallium scans, resulting in diagnosis.

FREIMANIS: Would you comment on gallium scans?

WISTOW: Certainly. I chose not to bring gallium into the discussion because it had been discussed before and also because our department does very few gallium studies, maybe one a month. This is obviously different from what other people do and it is due to a difference of opinion as to efficacy. We have chosen to use the liver-lung scan in nuclear medicine as our conventional study. It is something that can be done very quickly. Also, it just doesn't happen to be very effective and so we are actually going away from that. We see very few cases coming for subphrenics. Perhaps Dr. Wenzel has more experience with subphrenics and gallium. I do not negate the usefulness of gallium scans at all.

WENZEL: We had one presumed false positive subphrenic abscess scan. It turned out that there was a thin plaque of inflammatory tissue below the diaphragm. I always like to link together the gallium scan with some sort of other study—we do not have a body scanner right now—so we use ultrasound. I think ultrasound is really very effective to link those two together. In the diagnosis of any kind of a *positive* gallium scan in the abdomen—the gallium is really a sensitive agent. I like to separate the sensitivity and specificity in finding these inflammatory processes. But whether or not it is an abscess and should be drained or is just a focus of inflammatory process which is counteracted with antibiotics, can be determined by ultrasound. This is the way we like to approach subphrenic inflammatory processes in our department. Unless the lesion is in the liver, the liver scan has not been particularly helpful to me.

FREIMANIS: In our experience also, the *combination* of ultrasound and gallium scans looking for abscesses has been very helpful. There *is* still some value in conventional radiographic examinations as well. It is remarkable how much one can learn about subphrenic abscesses and intraabdominal abscesses when one goes back to conventional roentgen films in the questionable case.

LYONS: We had a really interesting discussion yesterday about how much we are learning from these new modalities and are able to go back to the original radiographs and recognize what really was happening to them. This is really a very good point, Dr. Freimanis. The new techniques are teaching us more about evaluation of the original radiograph.

Ultrasonography of Abdominal Vessels

George Leopold

Since the advent of gray-scale scanning and improved resolution real-time techniques, ultrasonic study of the abdomen has become increasingly popular. Ultrasonographers, on the other hand, have been forced to cope with a bewildering assortment of organs and tubes which were previously invisible on the older, bistable scans. Although at first confusing, the multitude of vessels now being recognized have become valuable allies. In many cases they provide not only information about the vessels themselves, but often serve as convenient roadmaps from which much information about the adjacent organs and masses may be garnered.

For convenience of discussion, the arterial system and the two major venous systems will be discussed separately.

Arteries

Prior to gray-scale, the aorta was the only vessel that could consistently be demonstrated by ultrasonic scanning. Its major utility was in the diagnosis and follow-up of abdominal aortic aneurysms. As such, it was a valuable supplement to the contrast aortogram which, in the case of a large amount of intramural thrombus, could grossly underestimate the correct external dimension.

With the introduction of gray-scale, generalized acceptance of the 3.5 MHz transducer (as compared to the older 2.25 MHz transducer) and suspended respiration single sector techniques, a new era was begun. In addition to the aorta, careful scans now frequently demonstrate the origin of the celiac, superior mesenteric, and renal arteries. In transverse scan, the celiac axis has a characteristic wing-shaped appearance as it divides into its two major branches, the common hepatic and splenic arteries (Fig. 1). In sagittal scan, it may be seen arising from the anterior aspect of the aorta approximately 1 cm cephalad of the superior mesenteric artery (Fig. 2). The latter arises from the aorta and then turns abruptly caudad—often paralleling the aorta for a considerable distance. In transverse scan, it may be seen in cross section—situated just anterior to the aorta. It is characteristically surrounded by a dense collar of

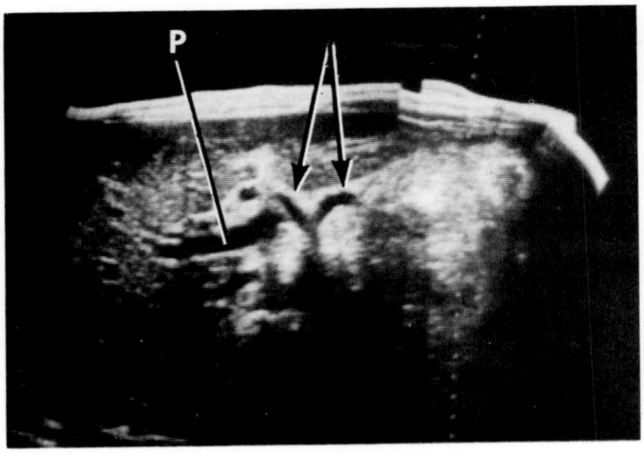

Figure 1. Transverse scan of the upper abdomen showing wing-shaped celiac axis (arrows) and portal vein (P).

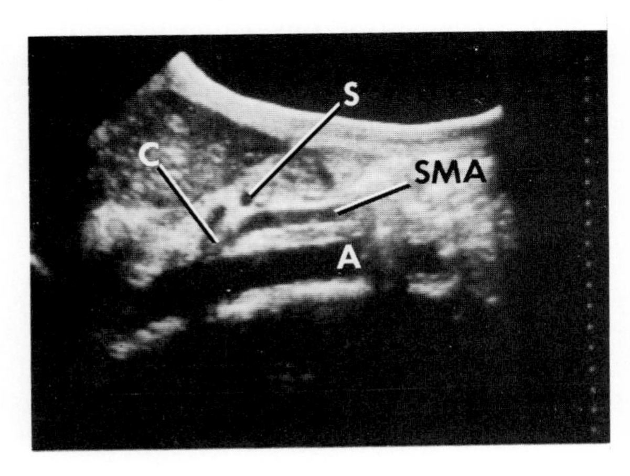

Figure 2. Midline sagittal scan. The aorta (A), celiac trunk (C), superior mesenteric artery (SMA), and splenic vein (S) are noted.

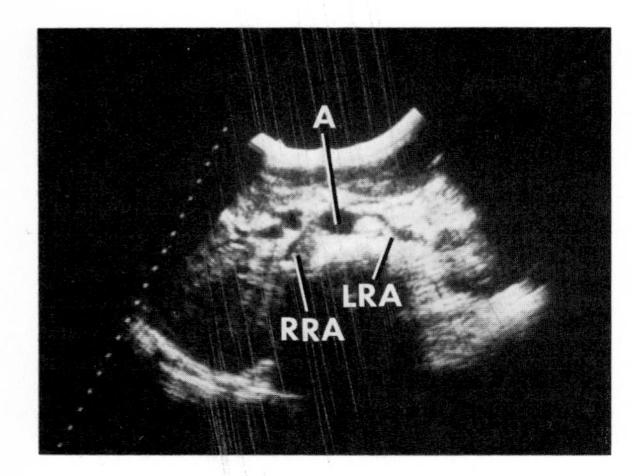

Figure 3. Transverse section showing the aorta (A) and both renal arteries (RRA, LRA).

echoes which are believed to arise from the fat in the root of the mesentery.

The renal arteries are seen arising from the lateral aspect of the aorta at a slightly lower level (Fig. 3). The right renal artery passes behind the inferior vena cava to reach the renal hilus. It is usually easily identified in cross section on longitudinal sections of the vena cava and often causes a distinct notch on the posterior surface of that vessel. The left renal artery is seldom seen on sagittal scans.

The inferior mesenteric artery is often small or occluded (older patients) and is infrequently seen on abdominal echograms.

The principle benefit of visualizing the branch arteries, in addition to detecting an occasional aneurysm, is observing deflection of them by a mass. Pancreatic head and body masses, for example, usually produce posterior displacement of the superior mesenteric artery, whereas paraaortic lymph node enlargement usually opens up the aortic-SMA angle. Less commonly, widening of this angle is produced by tumors arising in the uncinate process of the pancreas.

Systemic Veins

By utilizing suspended respiration, the systemic veins are caused to fill, and they become easy ultrasonic targets. The inferior vena cava is frequently larger than the abdominal aorta and is easily distinguished from that vessel by its ventral curve as it approaches the level of the diaphragm. Careful scanning adjacent to the inferior vena cava will disclose numerous venous branches extending back into the liver. These vessels may be distinguished from intrahepatic portal veins by their lack of prominent echoes at the margins of the vessel.

The right renal vein is a short, oblique extension of the inferior vena cava and may be seen on both transverse and slightly oblique longitudinal ultrasonograms. The left renal vein, on the other hand, is much longer and tortuous in its course. It is most easily identified on transverse scans where it may be seen arising from the left side of the vena cava, passing between the superior mesenteric artery and the aorta and then curving dorsally to enter the left renal hilus. In sagittal projection, it is often seen in cross section in the angle formed by the superior mesenteric artery and the aorta (Fig. 4). In this projection, however, it may easily be confused with a fluid-filled transverse portion of the duodenum which also has the same course.

The abdominal systemic veins are sensitive indicators of elevated pressures—usually due to right heart failure of any etiology. In such circumstances, the veins remain dilated, regardless of the phase of respiration. Such patients may frequently be referred because of liver function abnormalities, and the ultrasonographer therefore has an excellent opportunity to steer the workup in the proper direction. Tumors of the kidneys (and less commonly the liver) may invade the inferior vena cava. Tumor thrombus can usually be demonstrated ultrasonically and thus assist greatly in preoperative evaluation. Displacement of the inferior vena cava often occurs with lymph node masses on either the dorsal or ventral aspect of the vessel. Pancreatic head masses produce localized posterior displacement of the

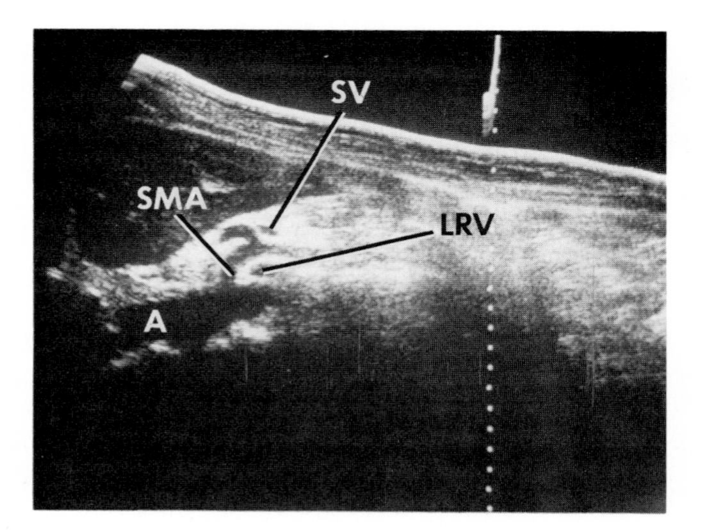

Figure 4. Sagittal midline section showing aorta (A), superior mesenteric artery (SMA), splenic vein (SV), and left renal vein (LRV). The vessel arising retrograde from the SMA is an accessory hepatic artery.

vena cava, while adrenal masses characteristically show anterior displacement.

Portal Venous System

Like the systemic veins, portal veins distend with suspension of respiration. Within the liver, numerous branches of the intrahepatic portal veins are normally seen. They are ordinarily marginated by dense echo-producing material which is believed to arise from accompanying smaller structures, such as the bile ducts, arteries, fat, and Glisson's capsule. With serial scans or real-time ultrasonography, they may be traced toward the liver hilus, rather than to the diaphragm. Portal venous radicles are seen on both transverse and sagittal scans of the liver. The extrahepatic portal vein, along with the other major tubular systems of the lesser omentum (gastrohepatic ligament), is best seen on oblique scans which parallel the angle of insertion of these vessels into the liver—frequently about 45°. In sagittal scan, the extrahepatic portal vein lies directly anterior to and in contact with the inferior vena cava. This is the site frequently chosen for surgical porta caval shunt.

The splenic vein is the transverse continuation of the portal vein across the abdomen and is therefore most easily appreciated on transverse scans (Fig. 5). It may be distinguished from the transversely running left renal vein by its position anterior to the superior mesenteric artery. On sagittal midline scan, it is seen in cross section as a small circle lying between the left lobe of the liver and the superior mesenteric artery.

The superior mesenteric vein is the largest tributary

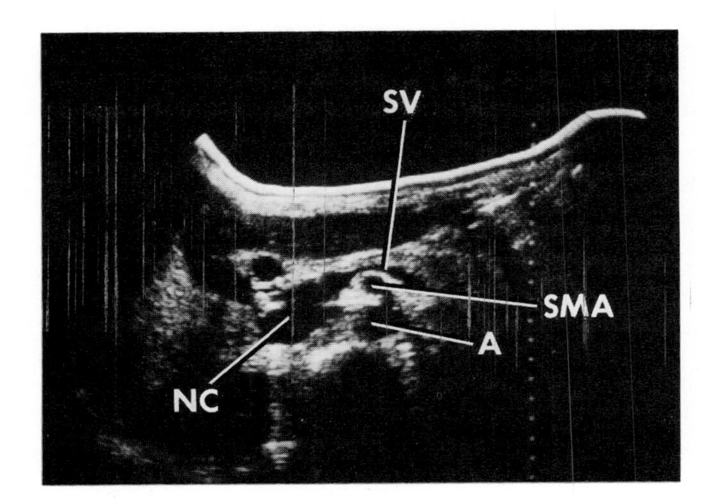

Figure 5. Transverse section, post porta caval anastamosis. The aorta, superior mesenteric artery, splenic vein, and inferior vena cava (NC), as well as the anastamosis are demonstrated.

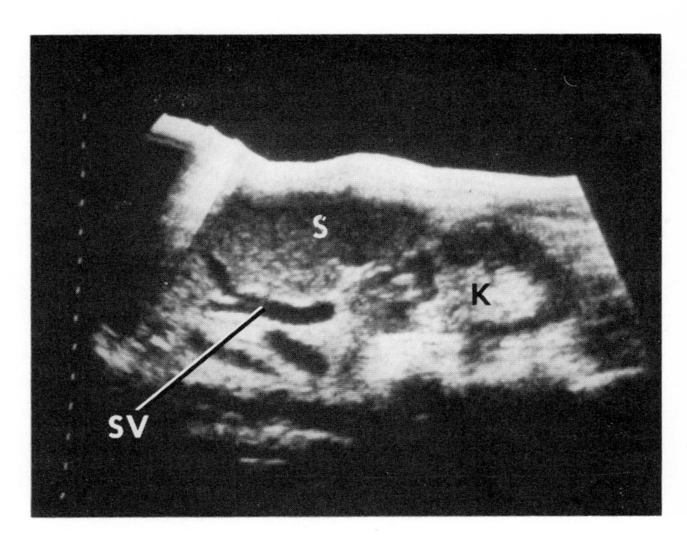

Figure 6. Sagittal section of enlarged spleen (S) and left kidney (K). The dilated splenic vein (SV) is seen exiting the splenic hilus.

of the portal venous system. In sagittal scan, it is usually noted anterior to and paralleling either the aorta or the inferior vena cava. It is distinguished from the superior mesenteric artery by its size (vein larger), by variation with phase of respiration, and by the fact that its course is toward the undersurface of the liver—not the aorta. In transverse section, it will frequently be possible to see the SMA and SMV on the same scan. It will then be noted that the vein also is positioned slightly anterior to the artery. It is also apparent that the superior mesenteric vessels, while usually closely related to each other, will vary considerably with respect to the aorta. Their relative right–left position is subject to considerable variation and is to some extent, at least, affected by the size of the liver.

Visualization of the portal venous system is important for a number of reasons. Failure to see the normal intrahepatic branches is highly suggestive of diffuse parenchymal disease such as fatty metamorphosis or cirrhosis. Dilatation of the extrahepatic portal vein and its branches may be seen either in congestive failure or in portal hypertension. Isolated dilatation of the splenic vein is common in many patients with splenomegaly (increased flow through the spleen) (Fig. 6). The largest splenic vein seen by us has been a case of arteriovenous fistula at the level of the hilus of the spleen.

The portal venous system is also the primary marker for finding many other types of abdominal pathology. The portal vein itself, for example, is an excellent landmark for finding the proximal portion of the common bile duct. The superior mesenteric and splenic veins clearly indicate the position of the body and tail of the pancreas.

Although presently the study of the abdominal vessels has proved to be quite exciting, the end is not yet in sight for this technology. As real-time instruments improve, even more physiologic information about them is expected to become apparent. It now seems quite likely that by combining B scan imaging with range-gated Doppler ultrasonic examination, flow in these vessels can be accurately calculated in a noninvasive fashion. This exciting step is expected to accelerate even more the development of diagnostic ultrasound as an abdominal imaging tool.

References

1. Sample W: Techniques for Improved Delineation of Normal Anatomy of the Upper Abdomen and High Retroperitoneum with Gray Scale Ultrasound, Radiology 124:197, 1977
2. Leopold G: Gray Scale Ultrasonic Angiography of the Upper Abdomen, Radiology 117:665, 1975
3. Carlsen E, Filly R: Newer Ultrasonic Anatomy in the Upper Abdomen I, JCU 4:85, 1976
4. Filly R, Carlsen E: Newer Ultrasonic Anatomy in the Upper Abdomen II, JCU 4:91, 1976

Computed Tomography and Gray-Scale Ultrasonography of the Adrenal Gland

W. F. Sample

In the past, the adrenal gland has been difficult to image directly by noninvasive techniques. Normal adrenals could occasionally be visualized and enlargements above 3 cm could reliably be detected with infusion nephrotomography. With the advent of computed tomography (CT) and gray-scale ultrasonography (US), additional noninvasive techniques capable of visualizing the adrenal gland became possible. In this report, the normal anatomy of the transaxial approaches, pitfalls common to both modalities, reported diagnostic accuracies and the relative role of each modality in patients with adrenal disease will be discussed.

Normal Anatomy and Pitfalls

The right adrenal gland is situated cephalic to the right kidney anteriorly. More specific anatomic landmarks transaxially are the inferior vena cava anteriorly, the medial aspect of the right lobe of the liver laterally, and the right crus of the diaphragm medially (Figs. 1 and 2). In the transverse plane, the right adrenal gland has the shape of a Y, an inverted V, or an upside down and backward L. Although the transverse approach is very successful with US, longitudinal scans can also be obtained for three-dimensional view.

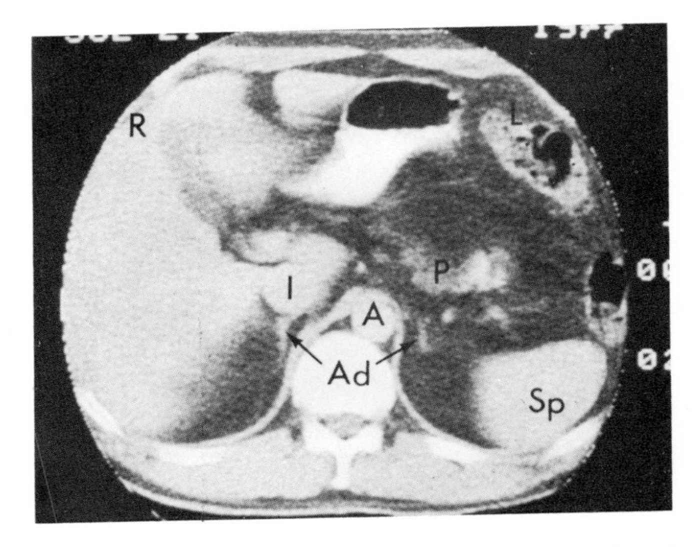

Figure 1. Computed tomogram through the adrenal glands showing the right adrenal gland posterior to the inferior vena cava (I) and the left adrenal gland lateral to the aorta (A). (R, right; L, left; P, pancreas; Sp, spleen; Ad, adrenal glands).

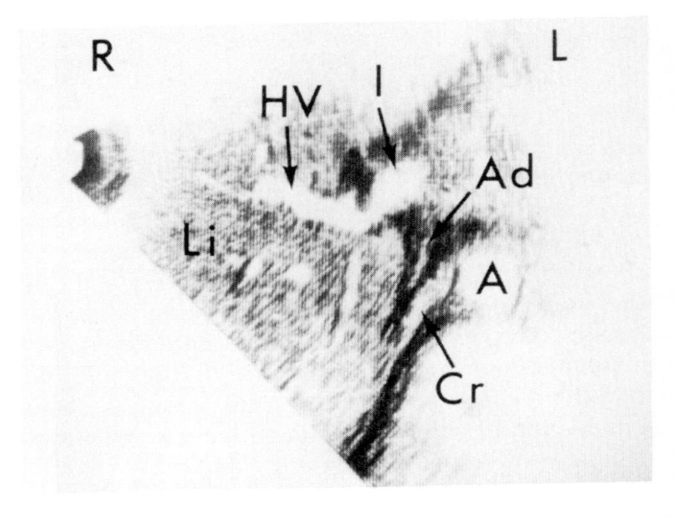

Figure 2. Transverse gray-scale sonogram through the right adrenal area showing the right adrenal gland (Ad) posterior to the inferior vena cava (I) and lateral to the crus of the diaphragm (Cr). (R, right; L, left; Li, liver; A, aorta HV, hepatic vein).

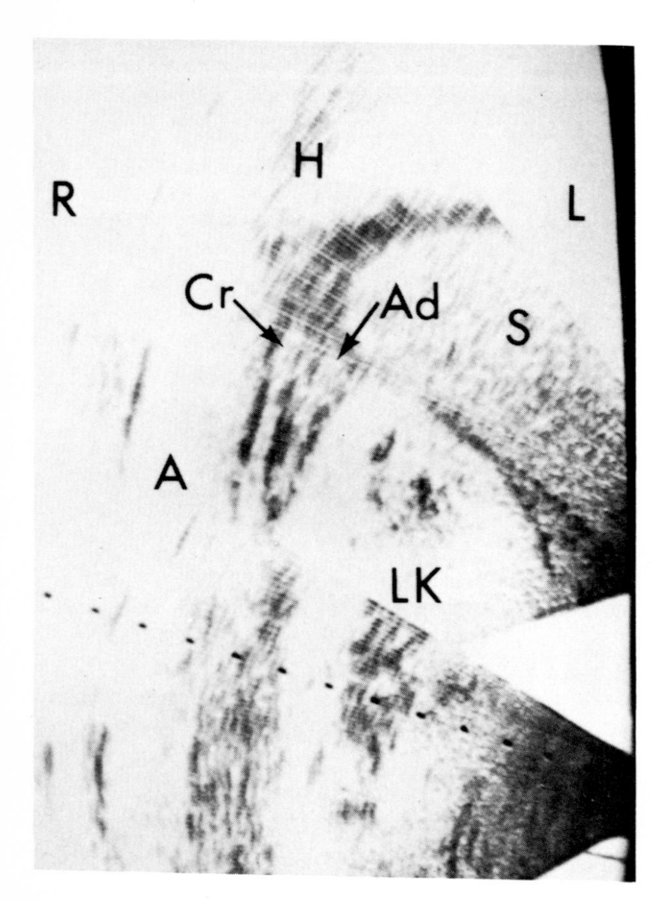

Figure 3. Longitudinal gray-scale sonogram through the left adrenal region showing the left adrenal (Ad) wedged between the left kidney (LK), the crus of the diaphragm (Cr), and the spleen (S). (R, right; L, left; H, head; A, aorta).

The left adrenal gland is situated on the anterior-superior and medial aspect of the left kidney. More specific boundary landmarks transaxially are the left crus of the diaphragm medially, the tail of the pancreas or the splenic artery anteriorly, and the spleen posterolaterally (Fig. 1). Although the left adrenal gland can occasionally be seen transversely with US, the gas in the stomach more commonly prevents visualization. A longitudinal oblique approach performed in the right side down decubitus position using the left kidney as a sonographic window can be used (Fig. 3).[1] With this technique, the adrenal gland has a triangular or crescent-shape wedge between the left crus of the diaphragm and the spleen.

Whether transverse or longitudinal approaches are used, certain pitfalls in identifying adrenal pathology are common to both techniques. The second and fourth portions of the duodenum are located at the same anteroposterior depth in the abdomen as the adrenal glands and can mimic adrenal pathology. Oral iodinated contrast is helpful in identifying the bowel with CT, while careful anatomic analysis plus the recognition of reflective mucous or shadowing gas is required for identification with US. The posterior-medial aspect of the fundus of the stomach or the esophageal gastric junction can also mimic left adrenal masses but can usually be identified as normal structures by a similar type of analysis.

The splenic artery is frequently tortuous in older patients and can be averaged in with the left adrenal gland giving the appearance of a mass on CT. With either modality the renal vessels can be prominent and masquerade as an adrenal mass, especially an extra-adrenal pheochromocytoma. Finally, the confluence of vessels at the aortic and inferior vena cava can be confused with a pheochromocytoma of the organ of Zuckerkandl.

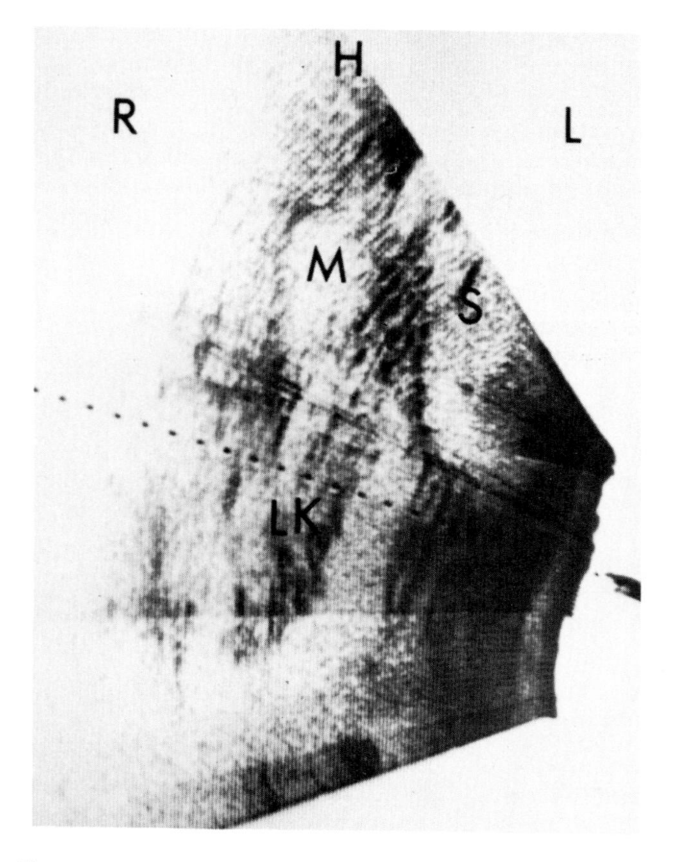

Figure 4. Longitudinal gray-scale sonogram through the left adrenal area showing a mass involving the adrenal gland (M) which at surgery proved to be an adenoma. (R, right; L, left; H, head; S, spleen; LK, left kidney).

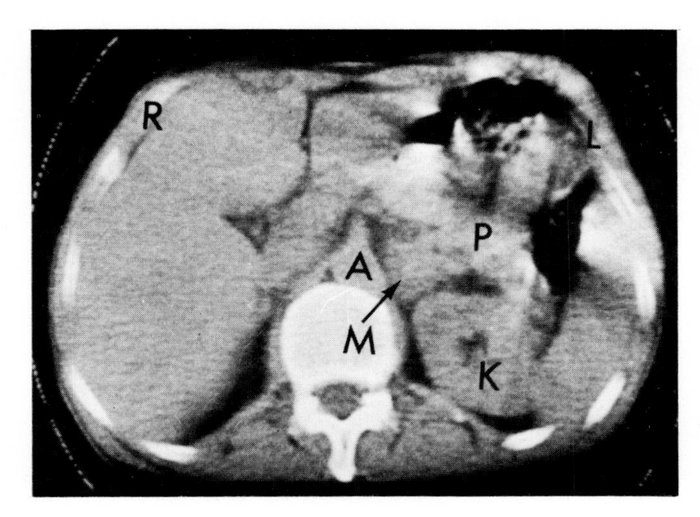

Figure 5. Computed tomogram through the left adrenal area showing an enlargement of the adrenal gland (M) which at surgery proved to be a pheochromocytoma. (R, right; L, left; P, pancreas; K, left kidney; A, aorta).

Diagnostic Accuracy of CT and US

Several investigators have reported diagnostic accuracies of 80 to 85 percent for lesions over 3 cm using bistable ultrasonic equipment.[2,3] My experience with newer gray-scale techniques has indicated a diagnostic accuracy of over 90 percent with tumors as small as 2 cm being identified (Fig. 4).[4]

The experience with CT of the adrenal gland is small but encouraging (Fig. 5).[5,6] Diagnostic accuracies exceeding 90 percent have been reported. In addition, the normal adrenal gland can be visualized in a high percentage of patients.

Relative Role of CT and US

With the variety of pathological processes that involve the adrenal gland, some of which cause only a functional rather than a size change, it can be anticipated that several imaging modalities will always be required for high diagnostic accuracy. CT and US should assume increasing roles in the screening of adrenal abnormalities which are likely to alter the size or shape of the gland. CT will have the advantage in more obese patients, whereas US may prove superior in thin patients. CT may be preferable initially in screening patients suspected of having a pheochromocytoma, since the potential extraadrenal locations are better visualized. The flexibility of ultrasound may be better in evaluating large masses in the high retroperitoneum of possible adrenal origin in that a vector type of analysis can be applied.[7]

References

1. Sample WF: A New Technique for the Evaluation of the Adrenal Gland with Gray-Scale Ultrasonography, Radiology 124:463, 1977
2. Davidson JK, Morley P, Hurley GD, Holford NGH: Adrenal Venography and Ultrasound in the Investigation of the Adrenal Gland: An Analysis of 58 Cases, Brit J Radiol 48:435, 1975
3. Kehlet H, Blichert-Toft M, Hancke S, Pedersen JF, Kristensen JK, Efsen F, Dige-Peterson H, Fogh J, Lockwood K, Hasner E: Comparative Study of Ultrasound, ^{131}I-19-Iodocholesterol Scintigraphy and Aortography in Localizing Adrenal Lesions, Brit Med J 2:665, 1976
4. Sample WF: Adrenal Ultrasonography, Radiology 127: 461, 1978
5. Sheedy PF, Hattery RR, Stephens DH, Williamson B, Brown LR, Hartmann GW: CT Scanning of the Adrenal Gland. Presented at the International Symposium and Course on Computed Tomography, Miami Beach, 1977
6. Sample WF: Computed Tomography and Gray-Scale Ultrasonography of the Adrenal Gland. Presented at the 63rd Scientific Assembly of the Radiologic Society of North America, Chicago, 1977
7. Whalen JP, Evans JA, Shanser J: Vector Principle in the Differential Diagnosis of Abdominal Masses: The Left Upper Quandrant. Am J Roentgenol 113:109, 1971

Discussion: Abdominal Imaging

Moderator: Thomas A. Verdon
Panelists: George Leopold
W. F. Sample

VERDON: Do either of you panelists have any specific recommendations as how to get rid of bowel gas or to move bowel gas around so that we can get an adequate study?

LEOPOLD: We generally move bowel gas by moving the patient to some other area of the radiology department for other studies. I really remain unconvinced that there is an efficient way to get rid of bowel gas despite what's in the literature. There are some papers which report a slightly statistically significant benefit from using simethicone prior to the examination. But it really doesn't work very well for us. When we consider the study to be undiagnostic, we do ask the patient to come back in a period of about 2 or 3 days. I must say that the percentage of patients who improve at the time of the second examination is remarkably small. If they are gassy the first time, they tend to be gassy the second time. We've tried all sorts of things. Someone recently suggested to me that smoking was a major factor and that if you could get the patient to cut down on smoking for a day or so before the examination, that may have some benefit. But the patient's sick and in the hospital and nervous anyway; you're not going to have any luck in getting him to stop smoking for that period of time. So, we don't use any special preparation. We do ask that the patient come in fasting for an upper abdominal study because we want to see the gallbladder at maximal size, but nothing else has turned out to be very effective in the evaluation of abdominal exams. People have talked about filling the bowel, or at least the stomach with fluid in an effort to visualize the body of the pancreas better. I think Dr. Freimanis was the first one who suggested that, but that technique has been relatively ineffective in our hands. I must say there is some recent work that looks sort of interesting: a newer Australian machine which we will be hearing more about shortly uses the prone position for scanning which places the body of the stomach at the low point; and they are currently investigating giving a diluted solution of methylcellulose to distend the stomach and provide a medium through which to scan, and hopefully, to see the body and the tail of the pancreas better than we presently do. I saw some pretty promising pictures. They combined that with an agent which is very similar to glucagon, but is unavailable in this country. It is a paralytic agent and immobilizes the gut for a brief period of time. They have found some interesting GI tract lesions by this technique. Our own attempts to give water, etc., to fill the stomach have been relatively ineffective except in proving that an observed mass in the left upper quadrant really may be the stomach. That can sometimes be a confusing feature. You see the left upper quadrant fluid-filled mass and you're not quite certain whether it's stomach or not. The easiest thing to do is just give them a swallow of tap water and watch with real-time ultrasound; the mass will fill with millions of bubbles that are swallowed, and that proves it's the stomach beyond a shadow of a doubt. But, usually all of that is pretty unsatisfactory. We really don't have a good way of eliminating bowel gas.

VERDON: Dr. Sample, do you have anything to add to that?

SAMPLE: No, I agree with everything Dr. Leopold says, 100 percent. Our philosophy is that we now have both modalities, CT and ultrasound, available to us and that we frequently go by body habitus. Some patients who come in to ultrasound not only are gassy, but have a generous body habitus and those we immediately triage to CT. If it's a real thin patient that's gassy, we continue to try and bring him back a couple of times because CT is often less than rewarding in a thin patient.

VERDON: Does anyone in the audience have any magic potions or remedies or suggestions? If not, as with the nuclear medicine gallium scan, the bowel continues to be a problem. Here's another question for Dr. Sample, it's in two parts. One, do you still consider the normal adrenal gland unresolvable with the 2.25 MHz transducer? And two, do you consider 8 cm to be the depth limit of a 3.5 MHz transducer in resolving the adrenal gland?

SAMPLE: Well, I think you have to talk about which adrenal gland you are looking at. You can resolve the right adrenal gland with a 2.25 MHz transducer because the axial resolution of that system is still very good and now that we've gone more to the transverse techniques that I showed you today, using those specific simple sector sweeps, where we've lined up the adrenal with the crux of the diaphragm. You can resolve the right adrenal with the 2.25 MHz transducer. The major failures that we have with the right adrenal gland is in people with a diseased liver, such as cirrhosis, where the sound waves cannot penetrate the liver adequately. Now, in dealing with the left adrenal gland, I still think you are primarily dealing with an 8-cm depth range and should use a 3.5 MHz transducer because you have a triangular shaped gland, the way you're approaching it most of the time, which means at least two of those three surfaces are relying on a lateral resolution of the system. Now the point is, if you don't have CT, sure, try. See the adrenal area and you will see moderate sized masses. Just realize that your resolution limits have changed from 1.5 to 2 cm to probably 3 to 4 cm, but in our hands, when we have CT available we quickly realize that we're going beyond 8 cm and we are getting the penetration, that this patient is going to be ideal for CT, and we are evolving more and more to study these patients with CT. In fact, one of the few areas that I still do the majority of scanning in ultrasound is in the adrenal area. It has been a very difficult area for teaching people to perform; I think it can be learned and can be successful, but I think that CT is going to replace ultrasound as the primary adrenal scanning procedure except with very thin patients. It's easy to perform and in most patients there are not many pitfalls.

LEOPOLD: Just one comment in relation to that question and a previous one. Many people, and I think Dr. Sample is an example of that, have considered in training technologists that there may be a considerable cross-fertilization in ultrasound and CT in terms of training. In fact, most of the technologists I've talked to who are accomplished ultrasound technologists are just bored out of their mind when asked to perform CT scans because there really isn't any comparison with the technological skills of the two procedures. It is not to say that one technique is better than the other. It is just to say that it is difficult to get people to interrelate to both.

VERDON: I have a question for Dr. Sample. In discussing the role of adrenal radionuclide scanning, can this procedure be performed without the use of a computer?

SAMPLE: I think it can be done without a computer but I think you are much more limited in the differential diagnoses you make, and I also think you lower your sensitivity in certain situations. Your ability to tell hyperplastic glands, nontumorous hyperplasia from normal glands, is probably lost without area of interest postcomputer processing type of analysis. However, picking up entirely functional adenomas, whether they may be causing the Cushings syndrome or hyperaldosteronism can be attempted. If you are trying to make the differentiation between macronodular hyperplasia and an abnormality causing aldosteronism, no, you're not going to be able to do it without computer processing. So once you realize your limitations, I think you can use radionuclide adrenal scanning effectively without a computer.

Pancreatic Imaging: Nuclear Medicine, Computed Tomography, and Ultrasound Correlation

Edward V. Staab
C. Leon Partain

Carcinoma of the pancreas is the fourth leading cause of cancer death in men and sixth in women. There has been a steady rise in the rate of death due to this malignancy for some time. Early diagnosis has not been possible and to date therapy has not altered the disease process. With only 2 to 5 percent of patients alive at 5 years, the pragmatic argument often voiced is why even try to diagnose pancreatic cancer with such a dismal outlook.

Abdominal pain is the most common presenting complaint of patients with pancreatic disease. The patient who is very sick with marked weight loss will usually undergo surgical exploration despite a myriad of negative test results. There is a large group of patients who have the above symptoms to a lesser degree, who may have pancreatic disease, that are not sick enough to warrant surgical exploration. If associated with weight loss and perhaps mildly abnormal liver-function studies, the possibility of pancreatic cancer is considered. It is in this group that the findings of the various diagnostic tests play a role in decision making of the primary physician.

Numerous tests to evaluate the pancreas have been developed. More recently, anatomical definition of the gland has been improved remarkably with the advent of gray-scale ultrasonography and computed tomography. Using these two modalities, most mass lesions can be detected.

What then is the role of pancreatic radionuclide imaging? Data are not available to compare the relative efficacy of the different modalities and to determine the answer to this question. In what sequence should the various procedures be obtained? As often happens in medicine, this decision is based on one's own experience and initial bias. Despite 15 years of experience, investigators still debate the efficacy of pancreatic nuclear imaging. On one side it is said that there are too many false positives to make it a useful screening procedure. Advocates suggest that very few false negatives will be experienced if interpretation is limited to just normal or abnormal glands.

We still use pancreatic imaging and feel that it is a useful technique. Pancreatic nuclear imaging is used for patients in whom the diagnosis is highly suspicious for pancreatic carcinoma and who have normal anatomical glands shown by ultrasound or computed tomography. Our diagnostic approach to the evaluation of suspected pancreatic mass is summarized in Figure 1.

Even though the prognosis for patients with pancreatic carcinoma is poor, we feel it is important to make a correct diagnosis because a patient should know why he hurts or is losing weight. Most patients and relatives can cope with serious illness if they know the cause of their symptoms. It is therefore important for a physician to make an accurate diagnosis. This also prevents multiple trips to different physicians and hospitals. Finally, with appropriate preparation, the adjustments following the loss of a relative may be easier.

This approach, however, has to be balanced against the need to make a diagnosis using the most simple, accurate, and inexpensive methodology.

Pancreatic Imaging by Ultrasound

Typical examples, using gray-scale ultrasound, of normal pancreas, pseudocyst, pancreatitis, and pancreatic carcinoma are shown in Figure 2. The value of ultrasound in imaging pancreatic lesions is well established.[1,2]

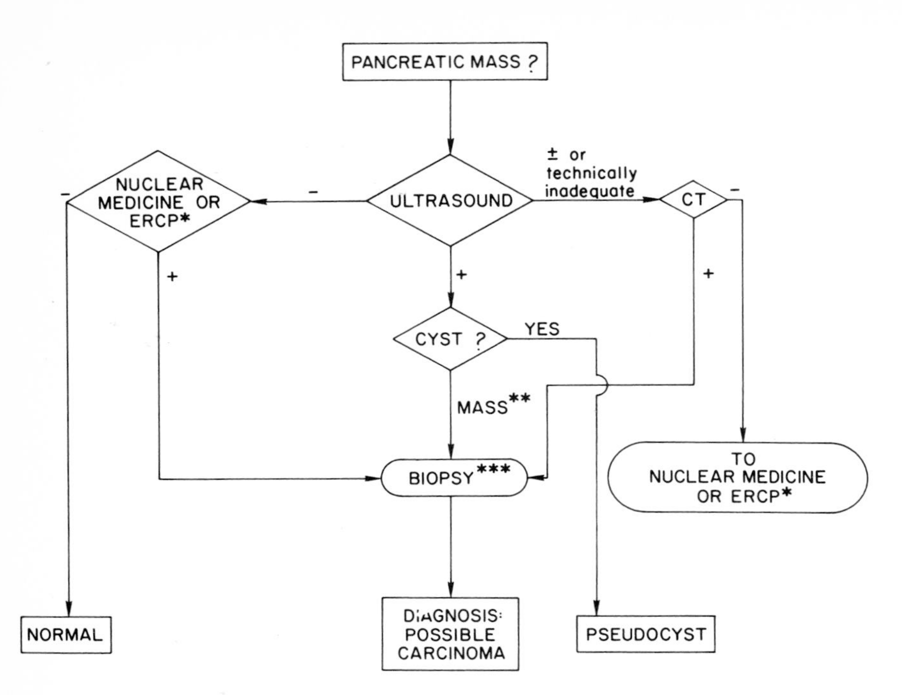

Figure 1. Diagnostic approach for suspected pancreatic mass.

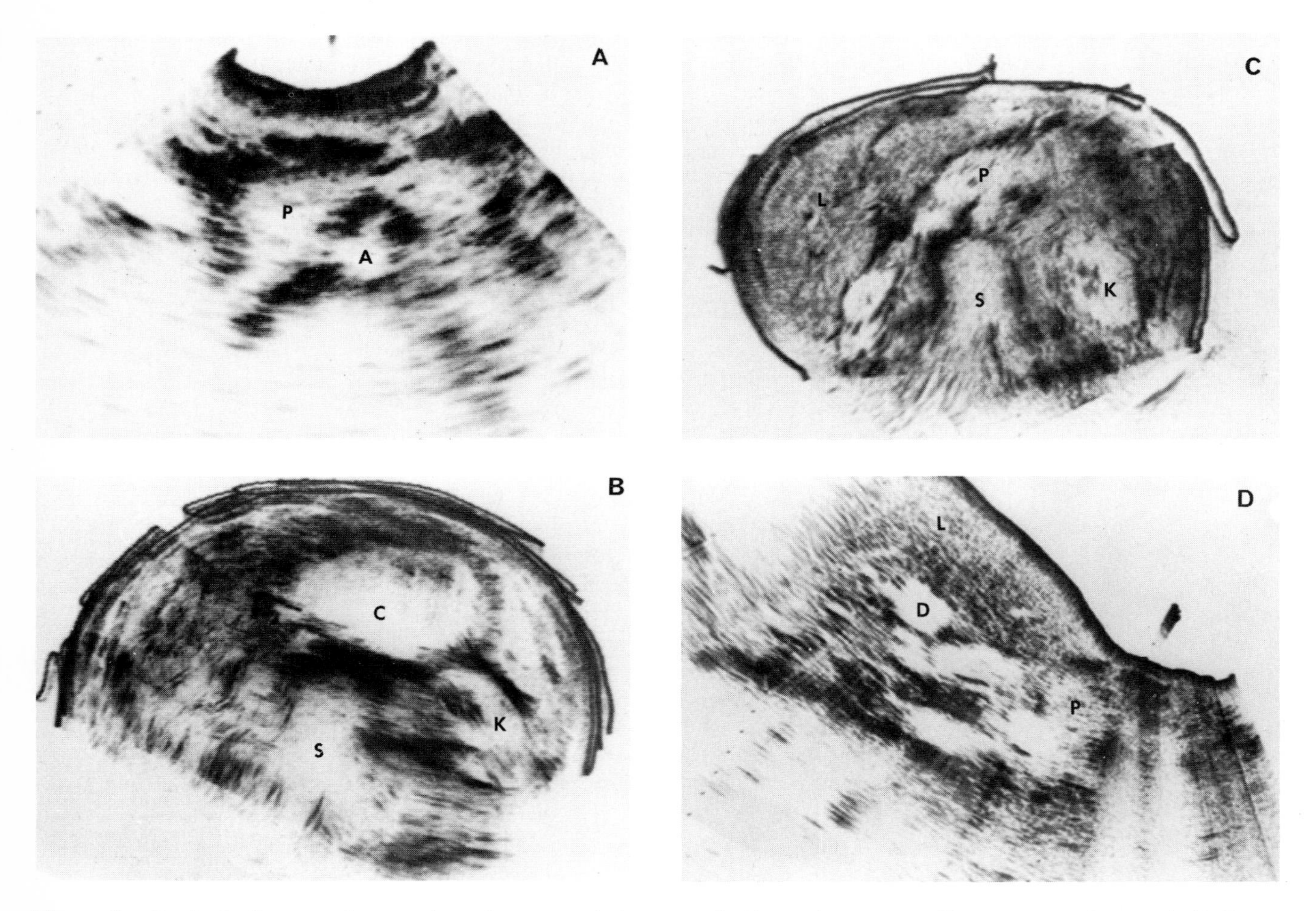

Figure 2. Typical ultrasound examples of pancreatic images: **A.** Normal pancreas (P, pancreas; A, aorta). **B.** Pseudocyst (C, cyst; K, kidney; S, spine). **C.** Pancreatitis (P, pancreas; S, spine; K, kidney; L, liver). **D.** Carcinoma (P, pancreas; D, dilated ducts; L, liver).

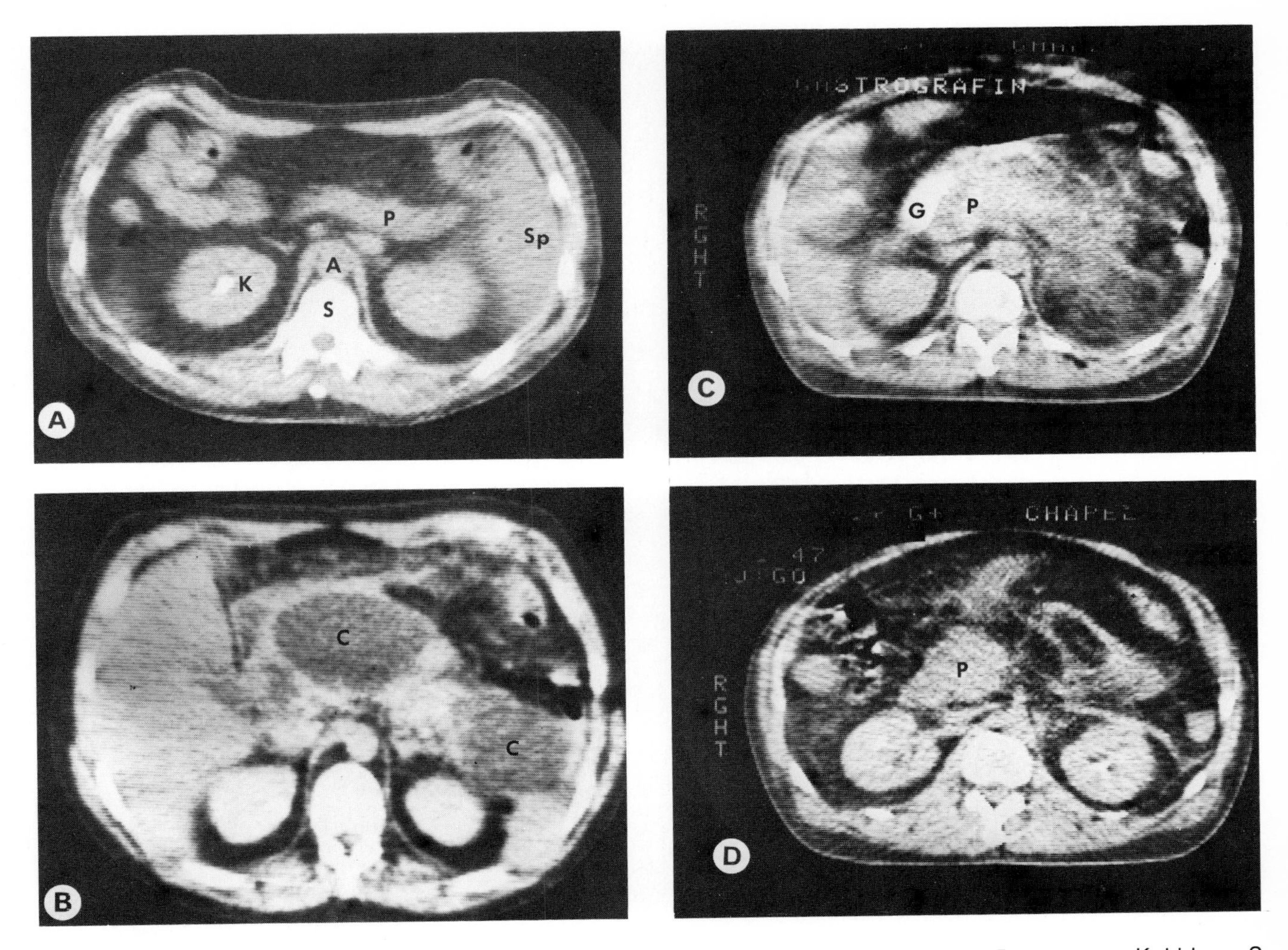

Figure 3. Typical transmission CT examples of pancreatic images: **A.** Normal pancreas (P, pancreas; K, kidney; S, spine; Sp, spleen; A, aorta). **B.** (C, pseudocysts). **C.** Pancreatitis (P, pancreas; G, gastrografin in duodenum). **D.** Carcinoma (P, pancreas).

Pancreatic Imaging by Transmission Computed Tomography

Typical examples, using an Ohio-Nuclear Delta (2.5 minutes per image) CT scanner, of normal pancreas, pseudocyst, pancreatitis, and pancreatic carcinoma are shown in Figure 3. Many investigators have demonstrated the efficacy of CT in pancreatic imaging.[3,4]

Pancreatic Imaging by Positron Emission Computed Tomography

Emission computed tomography using short-lived radionuclides is still an investigational technique.[5] There is hope that a pancreas-specific agent may be developed; promising candidates include carbon-11 tryptophan and carbon-11-valine.[6] An example of a normal pancreatic image using carbon-11 tryptophan is shown in Figure 4, generated by Oak Ridge Associated Universities.[7]

Pancreatic Imaging by Nuclear Medicine

The amino acid methionine is utilized by the pancreas in protein synthesis. From 5 to 10 percent of an injected dose localizes in the pancreas during the first hour after intravenous administration. The sulfur in methionine may be replaced by radioactive selenium-75 without any significant alteration in the biological behavior of the amino acid. Hence selenomethionine is an appropriate radiopharmaceutical for pancreatic imaging (see Table 1).

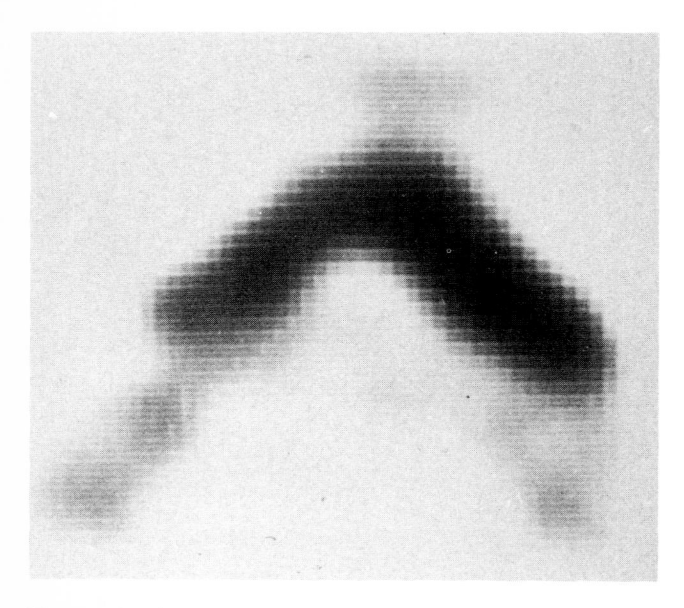

Figure 4. Normal pancreas by carbon-11 tryptophan, positron emitting, emission computed tomography.

Table 1. Characteristics of selenium-75 methionine

Radionuclide	^{75}Se
Physical half-life	120 days
Mode of decay	Electron capture
Principal gamma rays	136 Kev (57%)
	265 Kev (60%)
Decay product	^{75}As
Method of production	Neutron activation
Nuclear reaction	^{74}As (n, γ) ^{75}Se
Injection dose (IV)	250 μCi
Radiation dose (rads)	
Pancreas	3.0
Kidney	5.0
Spleen	4.0
Thyroid	1.5
Total body	2.0
Cost ($)	
Radiopharmaceutical	35.00 per patient dose
Total scan	205.00 per study

Methods

The simple gamma camera method involves the following.

1. There is an overnight or 12-hour fasting.
2. An oral meal of balanced fat, carbohydrate, and protein (some use 60 ml Lipomul) is given.
3. There is a 30-minute wait for gastric emptying.
4. Selenomethionine (maximum concentration of 20–30 minutes) is injected.
5. Patient is placed supine and the camera is angled approximately 25° cephalad and 15° toward the right shoulder so as to "look up under the liver." An experienced technologist will alter this depending on the configuration of the liver until the best possible profile of the pancreas is obtained.
6. Imaging begins immediately with 10-minute images (approximately 100,000 counts).

Selenomethionine distributes to numerous tissues. The largest concentration is in very metabolically active tissue where there is a high turnover of amino acids. The stimulated pancreas is one of these organs. There is less but significant concentration of the liver. Most patients can be satisfactorily positioned to separate the pancreas from the liver image. When the size and configuration of the liver is such that the two organ outlines cannot be separated by patient positioning, computer manipulation of data has been useful.

A second radiopharmaceutical is injected IV that localizes in the liver and not pancreas. By virtue of the difference in energies between technetium (140 Kev) and selenium (265 Kev), it is possible to electronically separate the activity in the different organs. Unfortunately, it is necessary to do the liver imaging first because of the lower energy of technetium. However, performing a liver scan is useful in patients with pancreatic cancer because of the high incidence of metastases to the liver.

Following a six-view liver scan, the patient is positioned to perform the pancreas study.

The computer-assisted technique is the following:

1. After liver scan, position patient as in the simple camera method to include the inferior border of liver.
2. Store 5 minutes of counts in the computer with spectrometer at 140-Kev range.
3. Change spectrometer to 265-Kev range.
4. Without moving the patient, inject 250 μCi of selenomethionine IV.
5. Fifteen minute images are taken off the camera oscilloscope and recorded on Polaroid or radiographic film.
6. Replay into computer memory stored counts from pancreas and liver (i.e., 265-Kev range).
7. Subtract the counts from the liver only (i.e., 140-Kev range) to give "best" picture of pancreas. This requires physician–computer interaction.

Interpretation

A normal pancreas may have several shapes, as illustrated in Figure 5. The pistol or dumbbell shapes are the most common. The tail of the pancreas may be congenitally absent, but this is very unusual. These different shapes are partially due to the angles at which

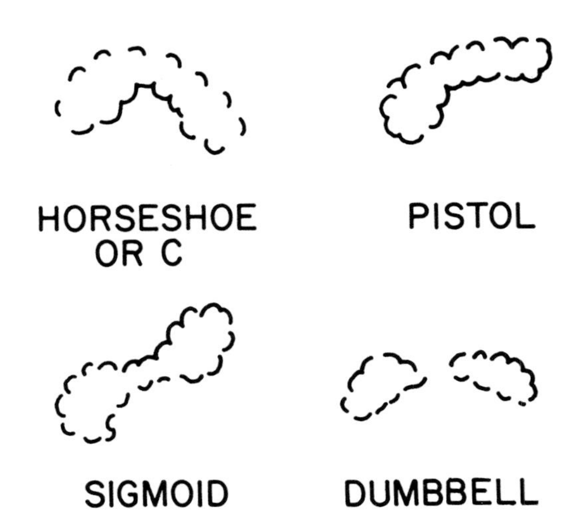

Figure 5. Various normal pancreas shapes.

the images are recorded. The pancreas appears thin in the body portion where it overlies the aorta and spine. Diffuse activity in the left upper quadrant represents bowel secretion and should not be confused with enlargement of the tail of the pancreas. A normal pancreas scan is shown in Figure 6.

The abnormal patterns include:

1. Nonvisualization.
2. Faint visualization.
3. Partial or segmental visualization.
4. Displaced pancreas.

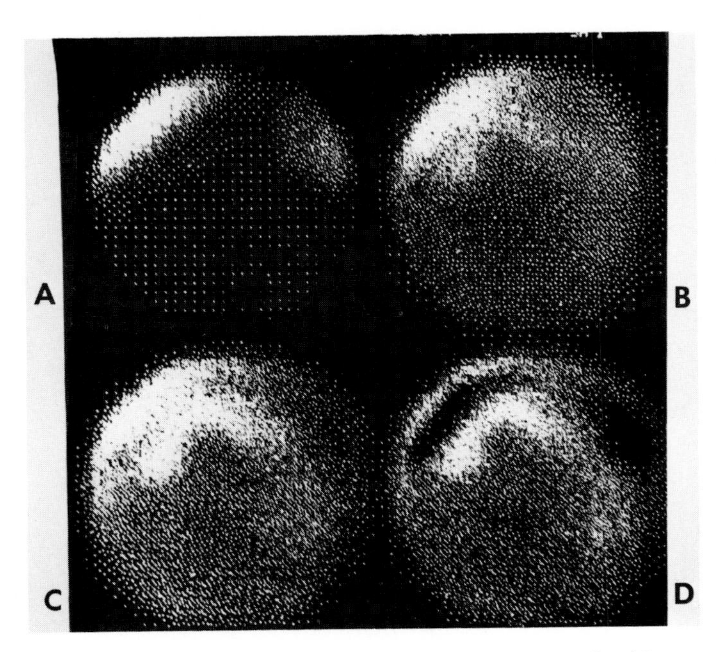

Figure 6. Normal nuclear medicine scan. **A.** Liver-spleen scan at top of camera. **B.** Liver-spleen and pancreas scan. **C.** Liver-spleen partially subtracted. **D.** Liver-spleen totally subtracted.

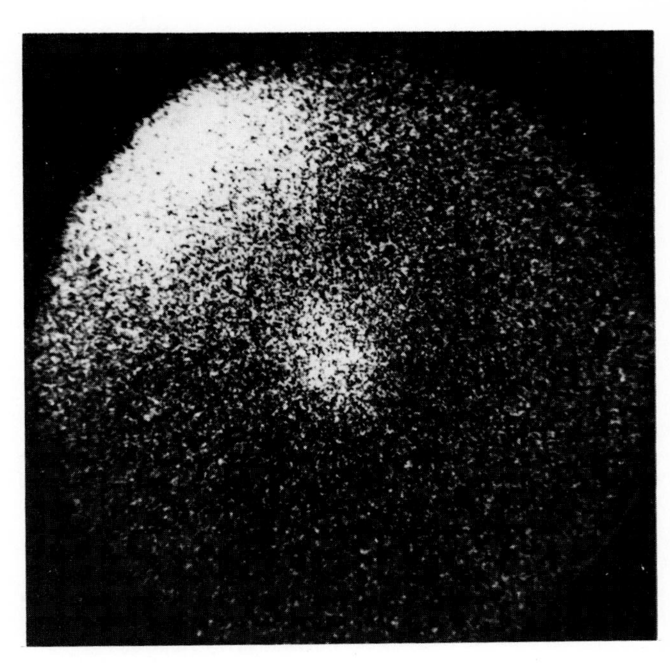

Figure 7. Abnormal pancreas scan, carcinoma in the tail of the pancreas.

An abnormal pancreatic scan is shown in Figure 7.

The most difficult pattern to interpret is the dumbbell shape. How much of a defect is allowed before this is considered abnormal? The computer-assisted study helps greatly with this problem.

Correlation

Data correlating nuclear medicine pancreatic imaging with computed tomography and/or ultrasound are not yet available. We have presented the diagnostic approach which is currently used at North Carolina Memorial Hospital (see Fig. 1.).

Conclusions

The presence of a normal anatomical gland that functions excludes pancreatic neoplasm with 90 to 95 percent certainty.[8–11] The future role of pancreatic nuclear medicine imaging will depend on analysis of current studies in light of the recent developments in gray-scale ultrasound and CT. Certainly the number of false positive studies will undoubtedly increase because the more easily diagnosed cases will no longer be included in the nuclear medicine imaging series.

Possible areas for improvement in pancreatic imaging include the addition of a quantitative dimension to the studies and perhaps the development of algorithms or diagnostic schemes to differentiate carcinoma from pancreatitis.

Also, the potential discovery of a pancreas-specific or a tumor-specific agent might make it possible to diagnose pancreatic carcinoma in a curable stage. Currently, however, pancreatic nuclear medicine imaging is most helpful in excluding pancreatic disease. A normal pancreatic scan is a highly reliable finding. Hence, the test is very sensitive but not at all specific.

References

1. Mittelstaedt C, Leopold GR: B-Scan Ultrasound of the Liver, Gallbladder, and Pancreas, Int Surg 62(5):277, 1977
2. Levitt RG, Geisse GG, Sagel SS, et al.: Complementary Use of Ultrasound and Computed Tomography in Studies of the Pancreas and Kidney, Radiology 126:149, 1978
3. Haaga JR, Alfidi RJ, Zelch MG, et al.: Computed Tomography of the Pancreas, Radiology 120:589, 1976
4. Haaga JR, Alfidi RJ, Harvilla TR, et al.: Definitive Role of CT Scanning of the Pancreas, Radiology 124:723, 1977
5. Phelps ME, Hoffman EJ, Mullani NA, et al.: Application of Annihilation Coincidence Detection to Transaxial Reconstruction Tomography, J Nucl Med 16(3):210, 1975
6. Washburn LC, Wieland BW, Sun TT, et al.: [1-^{11}C] DL-Valine, a Potential Pancreas-Imaging Agent, J Nucl Med 19:77, 1978
7. Emission Computerized Axial Tomography (ECAT) Project, Medical and Health Sciences Division, Oak Ridge Associated Universities, Oak Ridge, Tennessee, personal communication, 1978
8. Miale A, Rodriquez-Antunez A, Gill WM: Pancreatic Scanning after Ten Years, Sem Nucl Med 2(3):201, 1972
9. Miale A, Haynie TP: The Pancreas. In Clinical Scintillation Imaging, 2nd ed. New York, Grune & Stratton, 1975, p. 601
10. Landman S, Gottschalk A: Pancreas Scanning. In Diagnostic Nuclear Medicine. Baltimore, Williams & Wilkins, 1976, p. 456
11. Staab RE, Babb OA, Klatte EC, Brill AB: Pancreatic Radionuclide Imaging Using Electronic Subtraction Technique, Radiology 9(3):633, 1971

Gray-Scale Ultrasound and Computed Body Tomography of the Pancreas

W. F. Sample

For the first time in the history of imaging, noninvasive direct visualization of the glandular elements of the pancreas is possible. Previous noninvasive techniques, such as gastrointestinal x-rays, angiography, and ERCP, have detected the indirect effect of pancreatic abnormalities on surrounding or internal structures. With gray-scale ultrasound (US), the reflectivity of sound, and with computed tomography (CT), the attenuation of x-rays by the pancreas, images are distinct from those of the surrounding retroperitoneal fat. However, the limitations of each modality precludes visualization of the entire pancreas in every patient. Fortunately, the two modalities are complementary in that the pancreas is seen best in thin patients with ultrasound and the head and body are seen better than the tail. With computed body tomography, the pancreas is seen better in more obese patients irrespective of the gas pattern and the tail is frequently more adequately seen than the head.

Careful evaluations of pancreatic size have been reported for both imaging systems. Abnormalities can be detected by generalized enlargement of significant lobulations. However, the normal variations in pancreatic configuration and orientation make size analysis subject to errors. As a result, attention has been focused more recently on alterations in attenuation and graytone texture with computed body tomography and ultrasound, respectively.

Unfortunately, the attenuation values of pancreatic pathology may not be detectably different from the normal gland with the exception of pseudocyst formation. In contrast, the graytone texture differences of pancreatic abnormalities observed with ultrasound have been more consistent and can be detected prior to enlargement of the gland. However, the benign or malignant etiology of the process cannot be differentiated since the graytone texture patterns overlap.

During the past year, many reports confirming the diagnostic accuracy of ultrasound in the detection of pancreatic disease have appeared. The major problem continues to be incomplete visualization of the pancreas related to gaseous distention or obesity.

The present report describes the normal anatomy and variations of the pancreas as seen by CT and US, the diagnostic accuracy of CT and US for various pathologic entities, and the relative role of the two modalities in the evaluation of the patient with pancreatic disease.

Normal Anatomy

Both CT and US rely on similar anatomic landmarks for the identification of the pancreas.[1–7] In the region of the head of the pancreas, an anatomic sequence occurs from the patient's right to left including the gallbladder, the second portion of the duodenum, the head of the pancreas, the superior mesenteric vein and artery, and the fourth portion of the duodenum (Fig. 1). The inferior vena cava lies posterior to the head of the pancreas and the aorta is located deep to the superior mesenteric vessels. In general, CT and US equally visualize the gallbladder, the aorta, and the superior mesenteric artery. The second and fourth portions of the duodenum are better visualized by CT, especially after the addition of dilute iodinated oral contrast. The inferior vena cava and the superior mesenteric vein are more frequently identified with US. Alternative landmarks to the lateral aspect of the pancreas include the gastroduodenal artery anteriorly and the common bile duct posteriorly (Fig. 1). These two landmarks are potentially visible with either modality but are more frequently seen with US.

In spite of all of the anatomic landmarks, the head of the pancreas may not be well visualized with either modality. A lack of sufficient retroperitoneal fat may prevent adequate separation of the head of the pancreas from the surrounding structures with CT. Similarly, in obese or gassy patients, portions or all of the head of the pancreas may not be seen with US. If one took 100 patients at random, the head of the pancreas would probably be seen well in the same percentage of patients by each modality but not necessarily in the same patients. Fortunately, the two modalities are complementary in this region.

The neck or body of the pancreas is bounded by the superior mesenteric vessels posteriorly and the left lobe of the liver or the antrum of the stomach anteriorly. This region is probably equally well visualized by both modalities.

The tail of the pancreas is more variable in configuration and location. Anteriorly, it is bounded by the stomach and posteriorly by the left adrenal gland, the left kidney and/or the spleen (Fig. 2). In general, the tail of the pancreas is more frequently seen in its entirety with CT. The gas-filled body of the stomach often prevents visualization by anterior approaches with US. The administration of oral fluids will sometimes displace the gas allowing an ultrasonic window, but more frequently posterior scans through the left kidney are necessary. However, in neither case is the entire tail of the pancreas seen as well as with CT.

The glandular elements of the pancreas are usually isodense with the other organs of the upper abdomen on CT (with the exception of the liver). In contrast, the ultrasonic reflectivity is usually greater than the other surrounding organs and serves as an excellent background for the detection of disease.

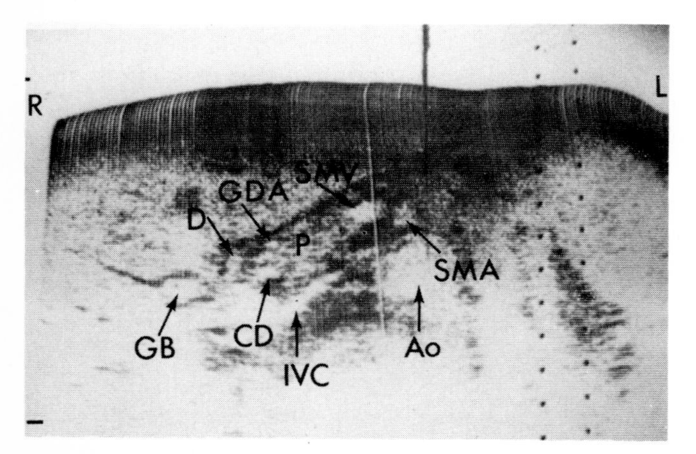

Figure 1. Transverse gray-scale sonogram through the head of the pancreas demonstrating the anatomic sequence from the patient's right (R) to left (L) (see text). (D, duodenum; P, head of the pancreas; GB, gallbladder; CD, common duct; Ao, aorta; IVC, inferior vena cava; GDA, gastroduodenal artery; SMV, superior mesenteric vein; SMA, superior mesenteric artery.)

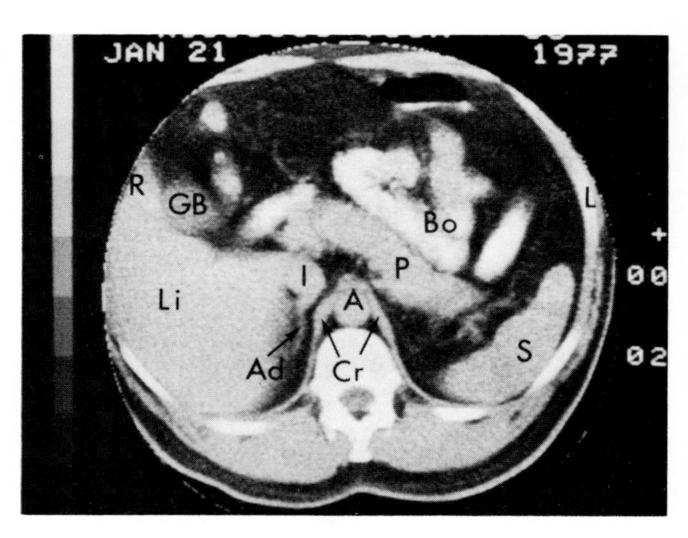

Figure 2. Computed tomogram through the body tail region of the pancreas (P). (R, right; L, left; I, inferior vena cava; A, aorta; S, spleen; GB, gallbladder; Li, liver; Ad, right adrenal gland; Cr, crus of the diaphragm; Bo, bowel.)

Diagnostic Accuracy of CT and US

Since diseases of the pancreas most frequently involve at least a part of the head and since CT and US visualize the head of the pancreas with equal incidence, it is not surprising that the reported general diagnostic accuracies for each modality are similar (80 to 90 percent).[8–18] However, a more detailed analysis of the limitations and pitfalls will help clarify the potential relative role of the two modalities in the diagnosis of pancreatic disease.

Pancreatic disease could be signaled by either a change in overall size, a localized enlargement, a texture change in the glandular elements without a mass effect, or the indirect obstructive effect on a recognizable ductal system. Any of these changes are potentially visible with CT and US. The specific nature of the pathologic process might be determined by characteristic transmission or reflective patterns with US and x-ray attenuation differences with CT.

A number of investigators have evaluated the normal size of the pancreas with CT and US and the results are similar.[19–21] Upper limits of normal for the anterior-posterior dimensions have ranged from: 2.5 to 3.5 cm for the head; 1.5 to 3.0 cm for the body; and 2.0 to 3.5 cm for the tail. In addition, three normal variants to the head-body-tail size relations have been identified: sausage shape, dumbbell shape, and a gradual tapering.

The texture and graytone features of ultrasound and the x-ray attenuation of CT by various pathologic processes involving the pancreas have also been evaluated. Both modalities reliably determine the cyst-solid nature

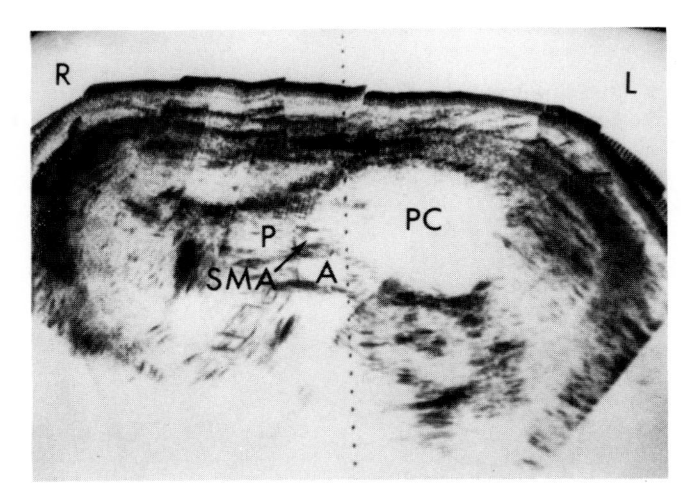

Figure 3. Transverse gray-scale sonogram through the pancreas demonstrating a pseudocyst (PC) in the body tail region of the pancreas. (R, right; L, left; P, head of pancreas; A, aorta; SMA, superior mesenteric artery.)

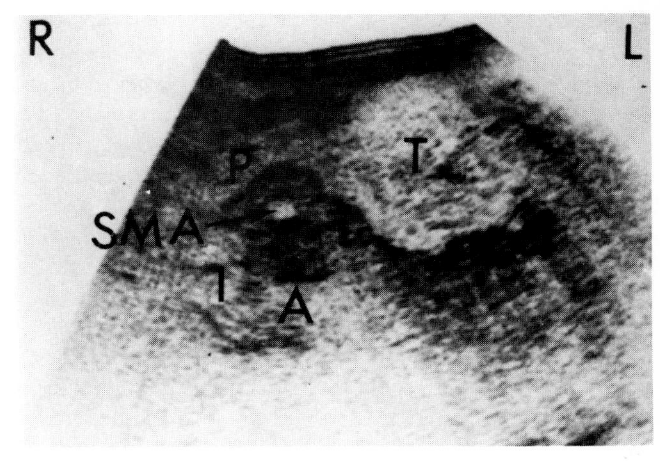

Figure 5. Transverse gray-scale sonogram through the pancreas demonstrating a solid mass in the body tail region (T) of the pancreas which, at surgery, was found to be a tumor. (R, right; L, left; P, head of pancreas; SMA, Superior mesenteric artery; I, interior vena; A, aorta.)

of an abnormality (Figs. 3 to 6). However, the attenuation values of various solid abnormalities have not always been different from normal pancreatic tissue making the detection of mass effects critical to CT. In contrast, most solid processes alter the normal ultrasonic texture and graytone of the pancreas, allowing for the detection of an abnormality without a mass effect. However, the changes in ultrasonic texture and graytone are not specific for benign or malignant processes.

A specific pathological process that may be better evaluated by CT is chronic pancreatitis. The subtle calcification sometimes accompanying this entity is better detected. Furthermore, the atrophy and fibrosis sometimes associated with chronic pancreatitis may not be visible with US and yet may be detectable with CT.

Relative Role of CT and US

The initial experience with CT and gray-scale US suggests that both modalities are going to be required for the detection of pancreatic disease. Neither imaging procedure is superior in all patients or all pathologic

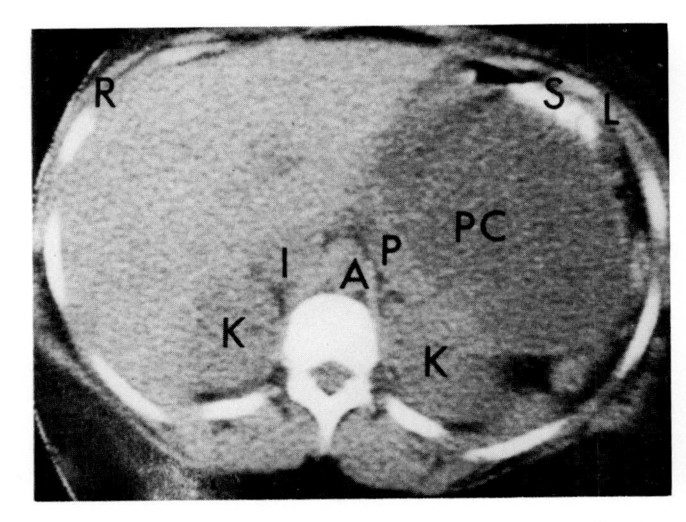

Figure 4. Computed tomogram of the same patient as in Figure 3 demonstrating the pseudocyst (PC) in the body tail region of the pancreas (P). (R, right; L, left; S, stomach; A, aorta; I, inferior vena cava; K, kidneys.)

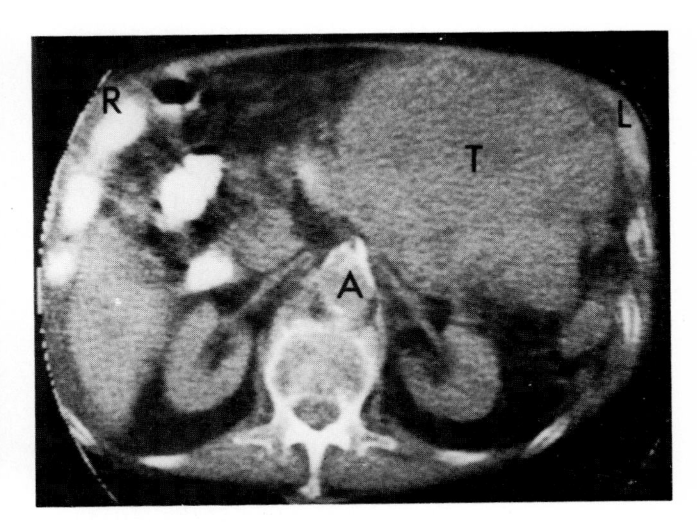

Figure 6. Computed tomogram of the same patient as in Figure 5 demonstrating the tumor mass (T) in the tail of the pancreas. (R, right; L, left; A, aorta.)

processes. Fortunately, the weaknesses of one are frequently offset by the strengths of the other. The choice of which procedure to use initially will depend upon availability and the experience of the imager. If both are available and the imager is knowledgeable in their use, then the choice will be tailored to each patient and the specific clinical problem. Since each modality evaluates different tissue parameters in some instances, the combined information can lead to more specific diagnoses.

References

1. Sample WF, Po JB, Gray RK, Cahill PJ: Gray-Scale Ultrasonography: Techniques in Pancreatic Scanning, Appl Radiol 4:63, 1975
2. Leopold GR: Echographic Study of the Pancreas, JAMA 232:287, 1975
3. Filly RA, Carlsen EN: Newer Ultrasonographic Anatomy in the Upper Abdomen: II. The Major Systemic Veins and Arteries with a Special Note on Localization of the Pancreas, J Clin Ultrasound 4:91, 1976
4. Ghorashi B, Rector WR: Gray-Scale Sonographic Anatomy of the Pancreas, J Clin Ultrasound 5:25, 1977
5. Sample WF: Techniques for Improved Delineation of Normal Anatomy of the Upper Abdomen and High Retroperitoneum with Gray-Scale Ultrasound, Radiology 124:197, 1977
6. Seidelmann FE, Cohen WN, Bryan PJ, Brown J: CT Demonstration of the Splenic Vein–Pancreatic Relationship: The Pseudodilated Pancreatic Duct, Am J Roentgenol 129:17, 1977
7. Kreel L: Computerized Tomography of the Pancreas. J Comp Axial Tomogr 1:287, 1977
8. Barkin J, Vining D, Miale A, Jr, et al.: Computerized Tomography, Diagnostic Ultrasound and Radionuclide Scanning: Comparison of Efficacy in Diagnosis of Pancreatic Carcinoma, JAMA 238:2040, 1977
9. Feinberg SB, Schreiber DR, Goodale R: Comparison of Ultrasound Pancreatic Scanning and Endoscopic Retrograde Cholangiopancreatograms: A Retrospective Study, J Clin Ultrasound 5:96, 1977
10. Lutz H, Petzoldt R, Fuchs HF: Ultrasonic Diagnosis of Chronic Pancreatitis, Acta Gastro-enterol Belg 39:458, 1976
11. Di Magno EP, Malagelada JR, Taylor WF, Go VLW: A Prospective Comparison of Current Diagnostic Tests for Pancreatic Cancer, New Engl J Med 297:737, 1977
12. Haaga JR, Alfidi RJ, Havrilla TR, et al.: Definitive Role of CT Scanning of the Pancreas, Radiology 124:723, 1977
13. Stanley RJ, Sagel SS, Levitt RG: Computed Tomographic Evaluation of the Pancreas, Radiology 124:715, 1977
14. Ponette E, Pringot J, Baert AL, et al.: Computerized Tomography and Ultrasonography in Pancreatitis, Acta Gastro-enterol Belg 39:402, 1976
15. Ferrucci JT Jr, Wittenberg J, Black EB, Kirkpatrick RH, Schaffer D: CT Findings in Disorders of the Pancreas. Exhibited at 63rd Scientific Assembly of the RSNA, Chicago, 1977
16. Kazam E, Katz R, Herbstman C, Behan M: Computed Tomography and Ultrasonography of the Pancreas—A Comparative Study. Presented at 63rd Scientific Assembly of RSNA, Chicago, 1977
17. Sheedy PF, Stephens DH, Hattery RR, MacCarty RL, Williamson B: Computed Tomography in Patients Suspected of Having Carcinoma of the Pancreas: Recent Experience. Presented at 63rd Scientific Assembly of RSNA, Chicago, 1977
18. Lawson TL: Sensitivity and Specificity of Pancreatic Ultrasonography. Presented at 63rd Scientific Assembly of RSNA, Chicago, 1977
19. Weill F, Schraub A, Eisenschler A, Bourgoin A: Ultrasonography of the Normal Pancreas, Radiology 123:423, 1977
20. Haber K, Freimanis AK, Asher WM: Demonstration and Dimensional Analysis of the Normal Pancreas with Gray-Scale Echography, Am J Roentgenol 126:624, 1976
21. Kreel L, Haertel M, Katz D: Computed Tomography of the Normal Pancreas, J Comp Assist Tomogr 1:290, 1977

The Biliary Tract: Diagnostic Evaluation by Ultrasonography

George Leopold

As recently as five years ago, most individuals working with diagnostic ultrasound believed that visualization of the gallbladder was seldom, if ever, possible. Rapid advances in technology have now altered the interpretation of these studies to the point that failure to visualize this organ is of itself a strong indicator of disease. Furthermore, the normal common bile duct can almost always be identified, and very minimal dilatation of the intrahepatic biliary radicles is usually apparent. These fortunate developments have resulted in greatly expanded use of the method for many disorders of the biliary tract.

Gallbladder

Technique

Due to the considerable variability of the normal contour of the gallbladder, it is difficult to describe a fixed routine which is satisfactory for all patients (Fig. 1). The patient should be fasting for at least 12 hours prior to the study. In general, one should begin with real-time examination to determine the long axis of the organ—a maneuver which may considerably speed up the procedure. If real-time is not available, routine transverse and sagittal scans are done over the right upper quadrant utilizing the single pass sector scan with suspended respiration. When the gallbladder is encountered, the scan interval is decreased to 0.5 cm or less. The examiner must be certain that sufficient scans have been done to include all of the gallbladder. Even if the gallbladder appears normal on the initial series of scans, several decubitus views are included to detect those calculi which are near the narrow neck of the organ and can easily be missed. If the gallbladder appears larger than normal, post-fatty meal scanning can be done to rule out obstruction.

Abnormalities

The primary use has been in the detection of calculous disease of the gallbladder. Numerous studies now demonstrate greater than 90 percent accuracy in the ability to determine the presence or absence of gallstones—even those that are no greater than 1 mm in diameter.

Stones are usually seen as dense echoes in the dependent portion of the gallbladder, but may occasionally float in a layer within the viscus (Fig. 2). Gravitational orientation of the stones is a helpful, reassuring sign. Many stones demonstrate a phenomenon known as acoustical shadowing, which is due to almost complete reflection and/or scattering of the ultrasound beam. Although not experimentally proven, it seems probable that the presence of shadow is more likely related to the size of the stone than to its composition.

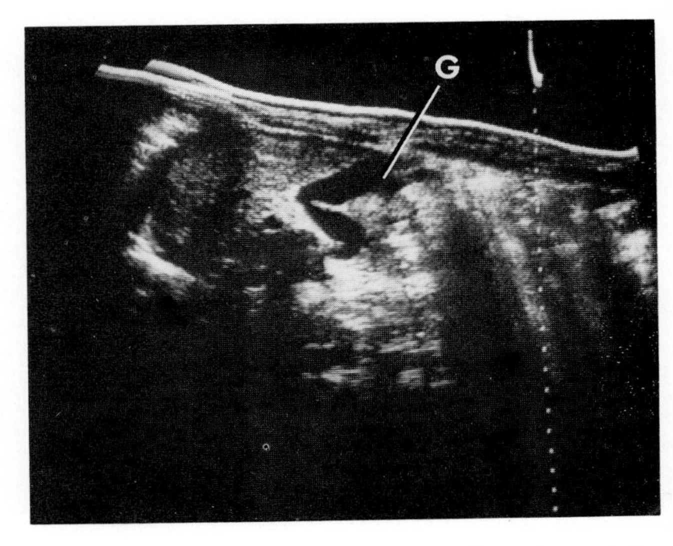

Figure 1. Sagittal scan, right of midline. One of the many possible variations in contour of the gallbladder (G).

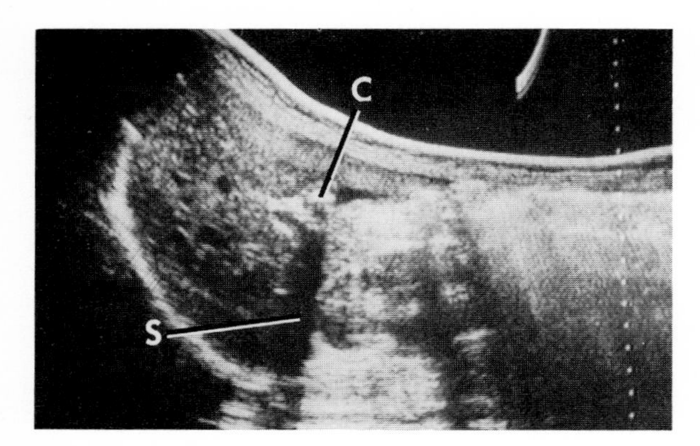

Figure 2. Sagittal scan showing a constricted gallbladder containing a calculus (C) and displaying a prominent acoustic shadow (S).

Thickening of the gallbladder wall is also a reliable indicator of disease (Fig. 3). It has not, however, been a consistent finding in our hands and, therefore, an apparent thin wall does not exclude the diagnosis of cholecystitis.

Most now agree that failure to visualize the gallbladder in a fasting patient is also strong presumptive evidence of a diseased gallbladder. At the time of surgery, such patients will usually have a small, contracted gallbladder full of stones and very little bile. It is the latter which is probably responsible for ultrasonic nonvisualization.

The accumulation of viscid bile within the gallbladder in patients who are fasting for long periods or have obstruction of the common duct provides a potential pitfall in diagnosis. This material, usually termed sludge, produces low level echoes and may be confused with gallstones. In general, sludge has a sharp upper

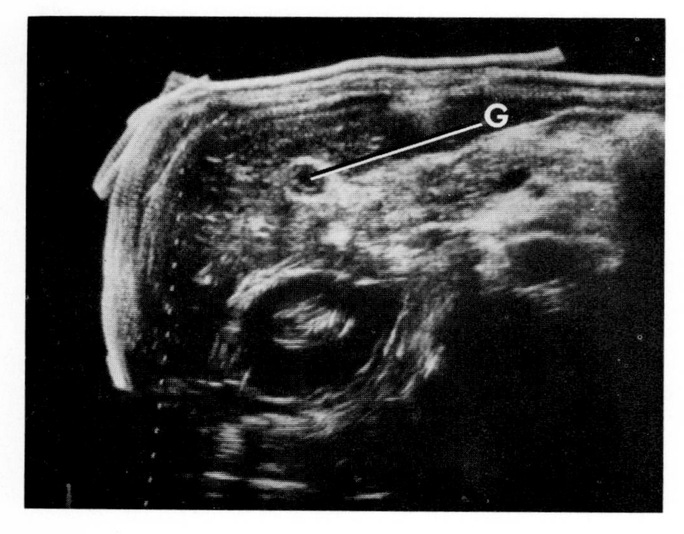

Figure 3. Transverse scan showing a small thick-walled gallbladder (G) in a patient with chronic cholecystitis.

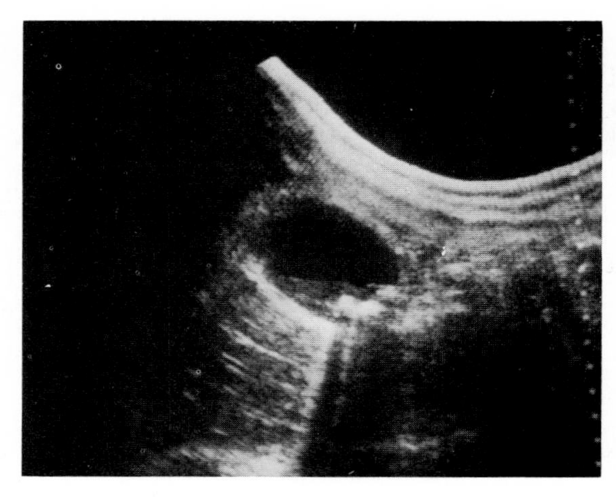

Figure 4. Transverse section of a diseased gallbladder containing both stones (dense echoes with acoustic shadows) as well as low density echoes from sludge.

margin (as opposed to the jagged contour of stones), and is composed of weak echoes (far less than that of the gallbladder wall) (Fig. 4). Decubitus views show that this material is much slower in assuming a dependent position in the gallbladder than stones. The precise relationship (if any) of sludge to gallstones remains to be proven.

The precise role of ultrasonography in calculous disease has not been established. Although the technique still occasionally misses small stones, we have seen a number of patients in whom stones were obvious on ultrasonography, but not seen on good oral cholecystograms. Certainly, any patient with long-standing, undiagnosed right upper quadrant pain would, therefore, be a candidate for this study. Likewise, the patient who presents to the emergency room with right upper quadrant pain is also an ideal candidate. Where previously intravenous cholangiography (2-hour test with morbidity and a 25 percent false negative rate for stones) would have been used, ultrasonography (15-minute, noninvasive, false negative less than 10 percent) now provides a better alternative.

Finally, in those patients with nonvisualization on oral cholecystography not related to gallbladder disease, ultrasound can often prove the gallbladder normal and at the same time demonstrate a liver or pancreatic disorder which might be the underlying problem.

Bile Ducts

Technique

For convenience, the biliary tree will be divided into three segments, each of which requires a different technique for demonstration.

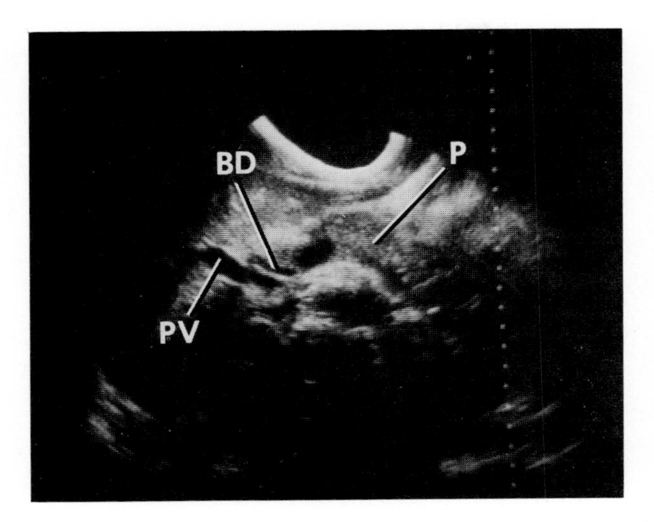

Figure 5. Oblique view showing a normal proximal common bile duct (BD) just anterior to the portal vein (PV). A segment of the pancreas (P) is also noted.

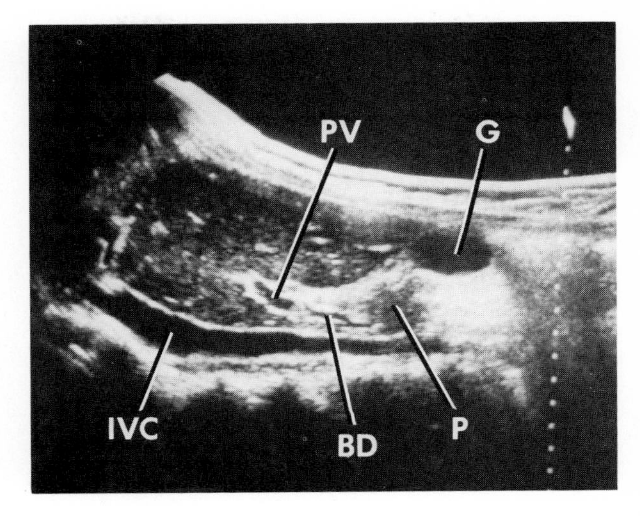

Figure 6. Sagittal scan along inferior vena cava (IVC) demonstrating the extrahepatic portal vein (PV), gallbladder (G), distal common bile duct (BD), and head of pancreas (P).

INTRAHEPATIC RADICLES The intrahepatic biliary radicles travel with the branches of the portal vein. Under normal circumstances, they are not visible on liver ultrasonograms. The major exception to this is the right hepatic duct which usually sits atop the right portal vein and is visible in sagittal scans of the liver in this area. The bile duct is much smaller than the portal vein at this point. Enlargement of the right hepatic duct is a sensitive indicator of the presence of obstruction.

When the more peripheral radicles dilate, they are easily distinguished from the normal hepatic vasculature by their wavy contour and by posterior acoustic enhancement. The latter occurs because bile, with its low protein content, absorbs much less of the incident ultrasound beam than does blood in the other tubular systems of the liver. If long-standing obstruction is present and bile becomes inspissated, this finding may become less obvious.

Dilatation of peripheral radicles is a relatively late phenomenon in obstructive jaundice and is not generally believed to occur prior to a week or more of complete obstruction of the more proximal ducts.

COMMON HEPATIC DUCT, PROXIMAL COMMON BILE DUCT These segments are closely related to the anterior aspect of the portal vein just prior to its insertion into the liver. It is therefore advantageous to determine the angle of insertion and then make a number of closely placed sections along that same oblique plane. If real-time equipment is available, this angle is usually readily apparent. Otherwise, numerous scans must be made with the trial and error method utilizing the contact scanner. We have found oblique scanning in left lateral decubitus position, as advocated by Kazam, to be quite helpful. This maneuver causes the air-filled bowel to drop away from the porta hepatis and positions a large segment of the transsonic liver in front of the bile ducts. Using the same position, it is sometimes possible to demonstrate the junction of the cystic and common hepatic ducts (as well as the proximal portion of the common bile duct)(Fig. 5).

DISTAL COMMON BILE DUCT After traveling with the hepatic artery and portal vein, the distal portion of the common bile duct angles posteriorly and caudally, passing behind the duodenum to enter the head of the pancreas (Fig. 6). This portion of the duct is usually quite close to the plane of the inferior vena cava, and scans along the cava frequently will demonstrate a long segment of it. Slight adjustments to the right or left may be necessary for maximum visualization.

Abnormalities

Analysis of the biliary tract has primarily been of benefit in detecting the presence of obstructing lesions usually of pancreatic origin. According to Sample, the upper limit of normal for the distal common bile duct is 6 mm in the patient who has not had previous biliary tract surgery. If surgery has been done, 10 mm is allowed (Fig. 7). It should be noted that these measurements do not relate to the usually accepted radiographic figures, since the latter are obtained under unphysiologic conditions and seldom take into account magnification factors.

In addition to dilatation of the biliary tree, sporadic

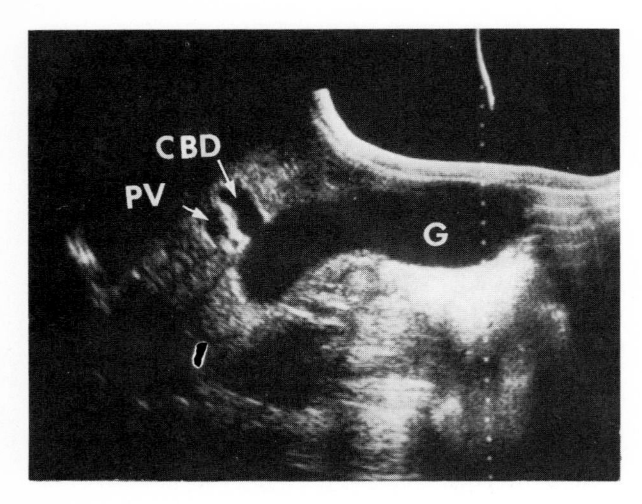

Figure 7. Sagittal scan showing a patient with a Courvoisier gallbladder (G), right portal vein (PV), and a segment of dilated common bile duct (10 mm) (CBD). This patient eventually proved to have pancreatic carcinoma.

reports are beginning to appear of choledocholithiasis and cholangiocarcinoma by the ultrasonographic method.

As experience with this technique increases, it is expected that its use will become even more widespread than at present.

References

1. Kazam E: Personal communication.
2. Sample W, Goldstein L, Kadell B, Weiner M: Gray-Scale Ultrasonography of the Jaundiced Patient—Presented at 63rd RSNA Meeting, Chicago, 1977
3. Leopold G, Amberg J, Gosink B, Mittelstaedt C: Gray-Scale Ultrasonic Cholecystography: A Comparison with Conventional Radiographic Techniques, Radiology 121:445, 1976

Gray-Scale Ultrasound and Computed Body Tomography of the Biliary Tree

W. F. Sample

The biliary tree has certain features which are particularly amenable to evaluation by computed tomography (CT) and ultrasonography (US). The fluid nature of the bile provides the contrast from surrounding tissues for US, whereas the lower density of the fatty constituents allows for detectable attenuation differences for CT. The present report will discuss the anatomy of the normal biliary tree as visualized by CT and US, pitfalls in making the differentiation of surgical versus medical jaundice, reported diagnostic accuracies for both modalities, and the potential relative role in the evaluation of the jaundiced patient.

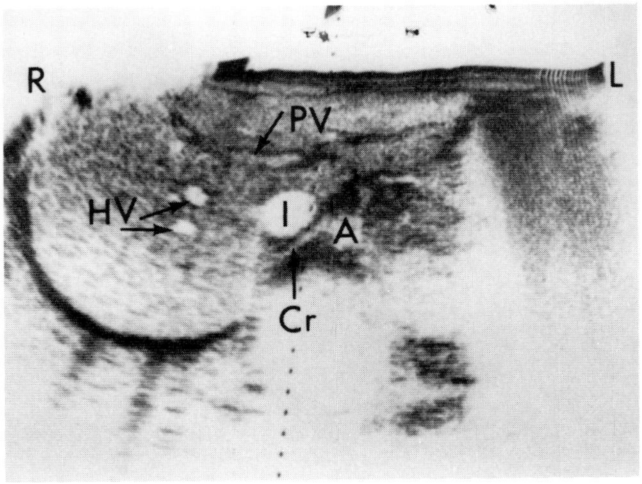

Figure 1. Transverse gray-scale sonogram of the liver showing the inferior vena cava (I) embedded in the caudate lobe and the normally visualized portal venous (PV) and hepatic venous (HV) systems. (R, right; L, left; A, aorta; Cr, crus of the diaphragm.)

Normal Anatomy

With the advent of CT and US, interest has been renewed in the transaxial tomographic anatomy of the biliary tree.[1–7] The intrahepatic portion of the normal biliary tree is below the resolution limits of both CT and US (Figs. 1 and 2). However, when the intrahepatic biliary system becomes even slightly dilated, the ducts are visible as additional fluid-filled structures on US or as low attenuation regions on CT (Figs. 3 and 4). With either modality, prominent portal venous radicals can be mistaken for dilated intrahepatic bile ducts. The

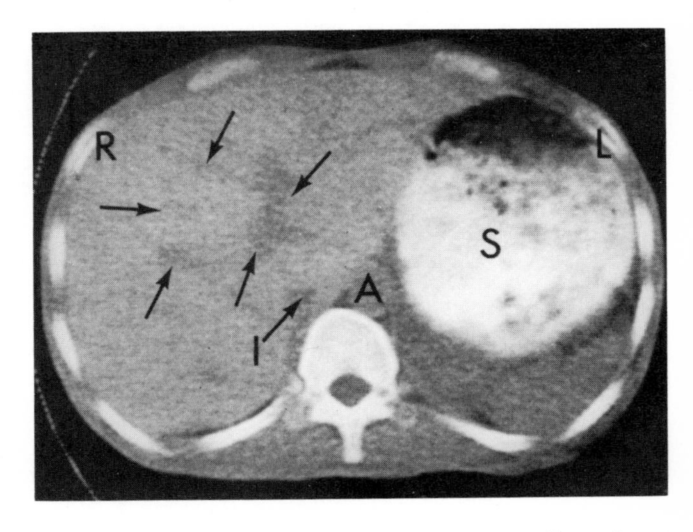

Figure 2. Computed tomogram through the liver showing the slightly lower density inferior vena cava (I) embedded in the caudate lobe of the liver and the aorta (A). The lower density regions within the liver indicated by the arrows are the same density as the inferior vena cava and the aorta and represent portal venous branches. (R, right; L, left; S, stomach.)

correct identification can be made on US by tracing the fluid areas back to their origin and on CT by carefully measuring the differential attenuation sometimes after intravenous contrast enhancement.

The extrahepatic biliary tree can be visualized normally without contrast enhancement by either modality in some patients. Usually the common hepatic or common bile duct will be detected in the anterolateral aspect of the hepatoduodenal ligament (Figs. 5 and 6). The location on US and the lower attenuation on CT will usually be sufficient for identification. Alternatively, with US, the common bile duct can be further traced to the posterior-lateral aspect of the head of the pancreas. Alternatively, intravenous cholegraphic contrast can be utilized to enhance visualization on CT.

Since the extrahepatic biliary tree can be seen normally in some patients, a critical question yet to be satisfactorily determined is the normal size of this portion of the biliary tree. Neither CT nor US require cholegraphic contrast to visualize the extrahepatic biliary system which probably distends the tree secondary to osmotic and physiologic mechanisms. In our own experience with US, the upper limits of normal for the diameter of the extrahepatic biliary tree is 6 mm.[8]

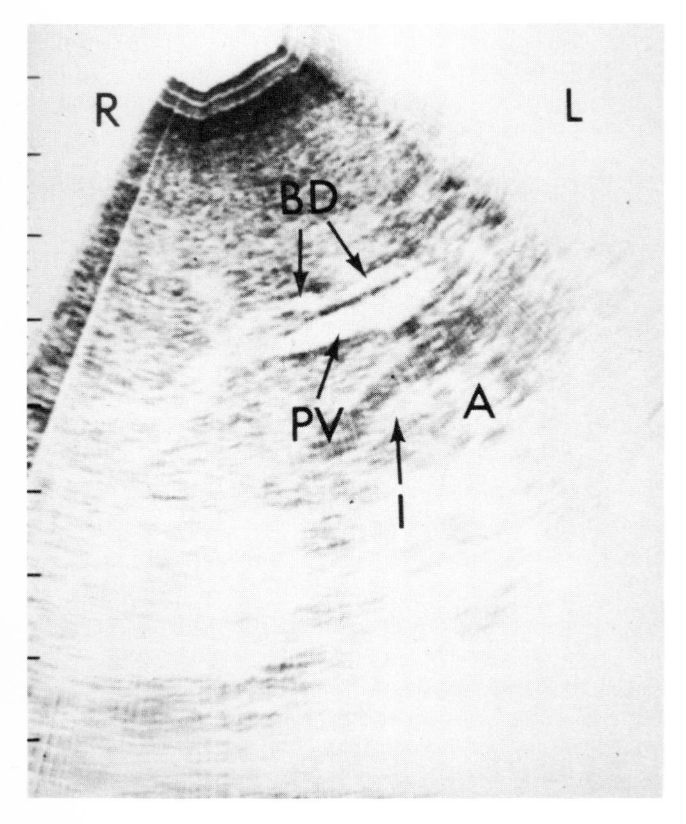

Figure 3. Transverse gray-scale sonogram of the liver showing a dilated bile duct (BD) anterior to the portal vein (PV). (R, right; L, left; I, inferior vena cava; A, aorta.)

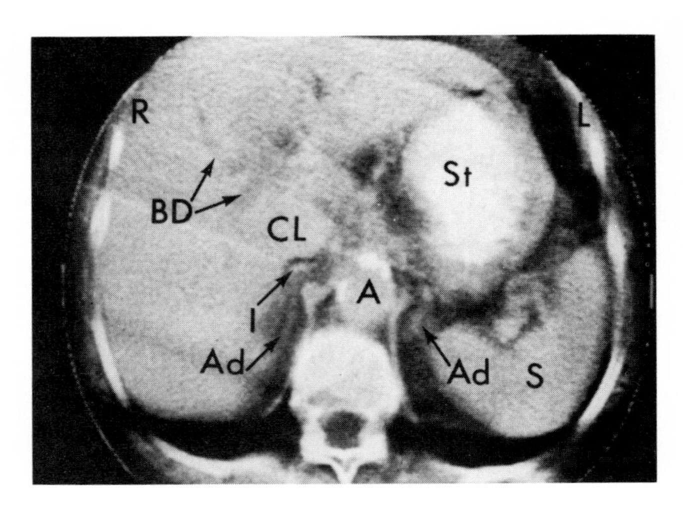

Figure 4. Computed tomogram through the liver showing dilated biliary ducts (BD) of lower density than the inferior vena cava (I) and the aorta (A). (R, right; L, left; CL, caudate lobe of the liver; St, stomach; Ad, adrenal glands; S, spleen.)

Diagnostic Accuracy of CT and US

Considerable experience has been reported with US in the evaluation of the jaundiced patient.[8–14] Nondiagnostic studies related to gas or bony impediments have occurred 5 to 10 percent of the time. Diagnostic accuracy in differentiating surgical from medical jaundice have ranged from 85 to 95 percent. In cases of surgical jaundice, the site and/or cause of the obstruction has been detectable 65 to 75 percent of the time (Fig. 7).

Experience is required to separate the normal or abnormal biliary system from nearby venous and arterial structures. Also, the size criteria for the normal extrahepatic biliary system without the magnification effects of x-ray and the osmotic effects of intravenous agents needs to be clarified. Although the distinction between a dilated and nondilated biliary system has been accurate, the nondilated abnormal biliary system remains a problem. Fortunately, this combination has not occurred frequently in the reported series.

Far less experience with CT and the jaundiced patient has been reported but the initial results are encouraging.[15–17] Nondiagnostic studies related to motion have occurred less than 5 percent of the time. Diagnostic accuracy in differentiating surgical from medical jaundice has been in the 90 to 95 percent range and the site and/or cause of obstruction has been visible 75 percent of the time (Fig. 8).

The major strength of both CT and US has been the sensitivity for detecting dilatation of the biliary tree, thereby signaling an obstruction which is likely a surgical condition. Both modalities are frequently capable

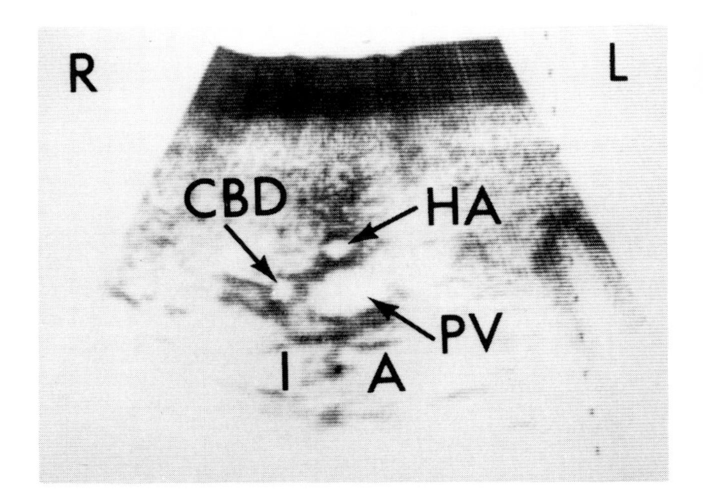

Figure 5. Transverse gray-scale sonogram through hepatoduodenal ligament showing the usual locations of the common bile duct (CBD), the hepatic artery (HA), and the portal vein (PV). (R, right; L, left; I, inferior vena cava; A, aorta.)

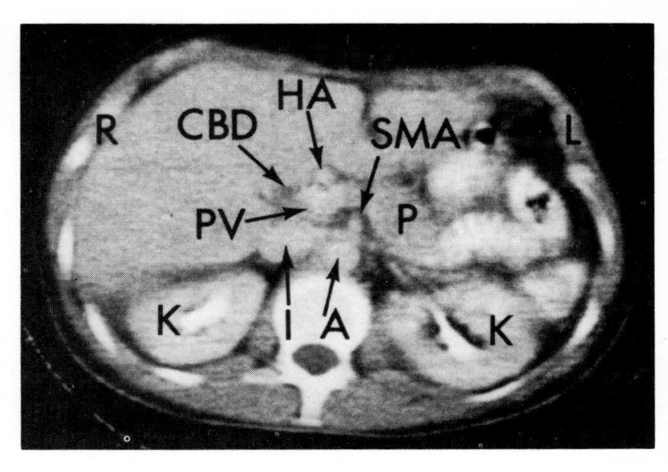

Figure 6. Computed tomogram through the hepatoduodenal ligament showing the usual location of the lower density common bile duct (CBD), the hepatic artery (HA), and the portal vein (PV). (R, right; L, left; K, kidneys; I, inferior vena cava; A, aorta; SMA, superior mesenteric artery; P, pancreas.)

of further visualizing the cause of the obstruction. CT has the advantage of not being hindered by bone or gas, whereas US has the advantage of flexibility and is less subject to errors generated by partial volume phenomena. Both modalities have potential difficulty with the nondilated biliary system involved with potentially surgical pathology. Fortunately, this form of surgical jaundice occurs infrequently and is usually encountered in cases with small common duct stones or sclerosing cholangitis.

Potential Role of CT and US in the Jaundiced Patient

In the jaundiced patient presenting acutely or incipiently, intravenous cholangiography has been valuable depending upon the degree of hyperbilirubinemia. Until the advent of US and more recently CT, a noninvasive method of accurately evaluating the jaundiced patient with serum bilirubins greater than 2–3 mg% was not possible. I-131 rose bengal scanning was fre-

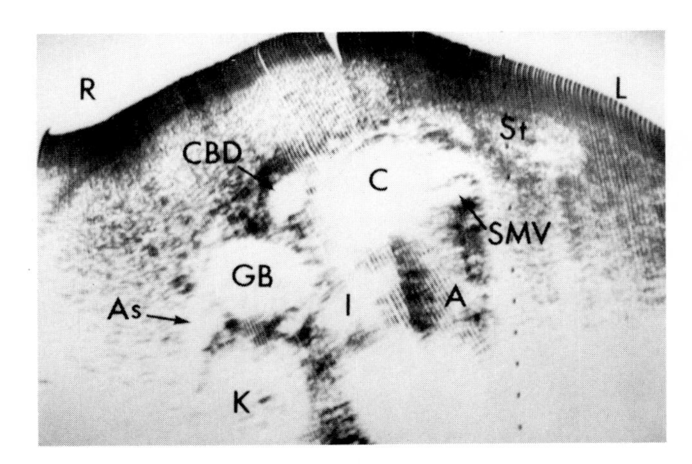

Figure 7. Transverse gray-scale sonogram through the head of the pancreas showing a pseudocyst (C) as the cause of the obstruction of the common bile duct (CBD). (R, right; L, left; I, inferior vena cava; A, aorta; K, right kidney; As, ascites in the hepatorenal angle; St, stomach; SMV, superior mesenteric vein; GB, gallbladder.)

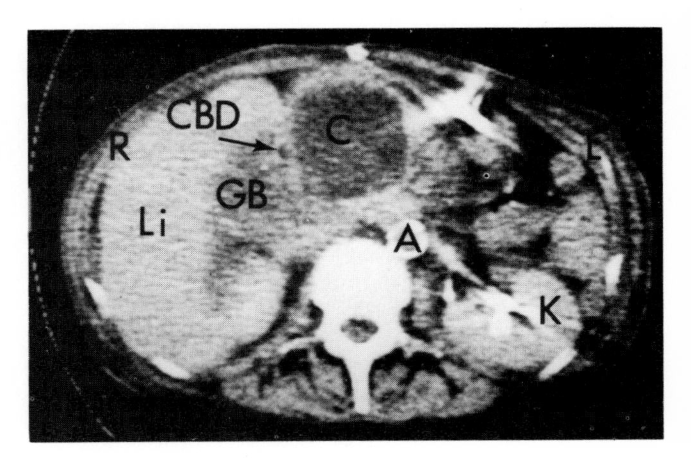

Figure 8. Computed tomogram of the same patient as shown in Figure 7 demonstrating the pseudocyst in the head of the pancreas (C) obstructing the common bile duct (CBD). (R, right; L, left; A, aorta; K, left kidney; GB, gallbladder; Li, liver.)

quently equivocal and the results of newer hepatobiliary scintigraphic agents have not been consistent.[18] If the accuracy data for US and CT in the jaundiced patient are further substantiated, it is likely that they both will be used to triage patients on to the most appropriate invasive procedure, i.e., transhepatic cholangiography, endoscopic retrograde pancreatico-biliary cholangiography, liver biopsy, or surgery.

With the diagnostic accuracy of both modalities similar for the jaundiced patient, the decision of which procedure to use initially will depend on ancillary factors such as the experience of the imaging physician with each modality and the body habitus of the patient. Hopefully, as more experience is gained with each modality, specific clinical situations where one modality is clearly superior to the other will be identified. Until that time, the existing data support the conclusion that either modality can be effective in experienced hands evaluating the jaundiced patient.

References

1. Filly RA, Carlsen EN: Newer Ultrasonographic Anatomy in the Upper Abdomen: II.The Major Systemic Veins and Arteries with a Special Note on Localization of the Pancreas, J Clin Ultrasound 4:91, 1977
2. Sample WF, Gray RK, Poe ND, Graham LS, Bennett LR: Nuclear Imaging, Tomographic Nuclear Imaging and Gray-Scale Ultrasound in the Evaluation of the Porta Hepatis, Radiology 122:773, 1977
3. Sample WF: Techniques for Improved Delineation of Normal Anatomy of the Upper Abdomen and High Retroperitoneum with Gray-Scale Ultrasound, Radiology 124:197, 1977
4. Sanders RC, Conrad MR, White RI, Jr: Normal and Abnormal Upper Abdominal Venous Structures as Seen by Ultrasound, Am J Roentgenol 128:657, 1977
5. Schapiro RL, Chin LC: A Primer in Computed Axial Tomographic Anatomy. 1. The Epigastrium, Comp Axial Tomogr 1:9, 1977
6. Lee TG, Henderson SC, Ehrlich R: Ultrasonic Diagnosis of Common Bile Duct Dilitation, Radiology 124:793, 1977
7. Kressel HY, Korobkin M, Goldberg HI, Moss AA: The Portal Venous Tree Simulating Dilated Biliary Ducts on Computed Tomography of the Liver, J Comp Assist Tomogr 1:169, 1977
8. Sample WF, Goldstein LI, Kadell B, Weiner M: Gray-Scale Ultrasonography of the Jaundiced Patient. Presented at 63rd Scientific Assembly of RSNA, Chicago, 1977
9. Goldberg BB: Ultrasonic Cholangiography: Gray-Scale B-scan Evaluation of the Common Bile Duct, Radiology 118:401, 1976
10. Isikoff MB, Diaconis JN: Ultrasound: A New Diagnostic Approach to the Jaundiced Patient, JAMA 238:221, 1977
11. Goldstein LI, Sample WF, Kadell BM, Weiner M: Gray-Scale Ultrasonography and Thin Needle Cholangiography in the Evaluation of the Jaundiced Patient, JAMA 238:1041, 1977
12. Malini S, Sabel J: Ultrasonography in Obstructive Jaundice, Radiology 123:429, 1977
13: Taylor KJW, Rosenfield AT: Grey-Scale Ultrasonography in the Differential Diagnosis of Jaundice, Arch Surg 112:820, 1977
14. Vicary FR, Cusick G, Shirley IM, Blackwell RJ: Ultrasound and Jaundice, Gut 18:161, 1977
15. Harell GS, Marshall WH, Jr, Breiman RS, Seppi EJ: Early Experience with the Varian Six-Second Body Scanner in the Diagnosis of Hepatobiliary Tract Disease, Radiology 123:355, 1977
16. Havrilla TR, Haaga JR, Alfidi RJ, Reich NE: Computed Tomography and Obstructive Biliary Disease, Am J Roentgenol 128:765, 1977
17. Levitt RG, Sagel SS, Stanley RJ, Jost RG: Accuracy of Computed Tomography of the Liver and Biliary Tract, Radiology 124:123, 1977
18. Ronai PM: Hepatobiliary Radiopharmaceuticals: Defining their Clinical Role will be a Galling Experience, J Nucl Med 18:488, 1977

Discussion: Gallbladder and Biliary Tract Imaging

Moderator: Thomas A. Verdon
Panelists: George Leopold
W. F. Sample

VERDON: What is the use of ultrasound for the emergency diagnosis of cholecystitis?

LEOPOLD: That's one area where its use has been the most clearcut in our experience and it is now well recognized by the house staff at our hospital as a primary procedure. When a patient comes into the emergency room with acute right upper quadrant pain and the surgeon is contemplating operating, if it is acute cholecystitis, that patient, who would previously have received an intravenous cholangiogram, is now sent directly to ultrasound, and if he has stones the surgeons are quite happy to go ahead and operate. I think that's an entirely logical way to procede because if you look at cholangiography, even with the presence of reasonable levels of bilirubin, and examine that study for cholelithiasis, there is about a 25 percent false negative rate. It's far too high to be used as an exclusive test to exclude the presence of gallstones. So our surgeons have been much happier to go to the ultrasound, which can be performed quickly and does not have any of the risks of intravenous cholangiography.

SAMPLE: This is certainly our experience, too. There are some hard data on this that were presented at AIUN this year of a large series by Dr. Molini where they examined cases of acute abdomen referred through the emergency room and found an 85 to 90 percent rate making this specific diagnosis. The other thing that you should be aware of that was alluded to earlier today is the high success rate of the newer radionuclide hepatobiliary agents at detecting an obstructed cystic duct. The only reason in my mind that the ultrasound study is preferred over that study is that if your radionuclide study is normal, it still may not tell you what is going on in that patient. It just tells you that the gallbladder isn't the disease problem. Furthermore, there may still be other pathology in the gallbladder that may be missed, such as a stone; it has been clearly demonstrated with even the newer hepatobiliary agents that it does not do well in chronic cholecystitis or in evaluation of patients with gallstones. We chose ultrasound because we can also make other diagnoses, like hydronephrosis or pancreatitis, etc., any of which can give this clinical picture of acute right upper abdominal pain without jaundice. Because of the flexibility, I think this is clearly the way to go. Why not CT? Because CT does nowhere near as well with stones in the gallbladder as ultrasound does, and this is probably the main cause for most people with this type of right upper quadrant pain. So, I think in the long run, ultrasound may turn out to be the major screening procedure in this group.

LEOPOLD: There is never any substitute for the clinical and physical examination. All of this information is very helpful. The problem is that there are still many things that can produce a similiar clinical situation. A perforated ulcer can be exquisitely tender right over the gallbladder. And I have to think that an imaging procedure which can see all of these areas reasonably well in a relatively high percentage of patients is the way to go.

VERDON: Is it possible to do any other kind of gallbladder function test with ultrasound other than just pure anatomy? Can you bring a person in in a fasting state and give them something to empty the gallbladder, and can you detect any type of functional abnormality by using ultrasound with this technique?

SAMPLE: A number of people have looked at either fatty meal administration or cholecystokinin as a part of the procedure in which all you find is an unusually large gallbladder but no evidence of gallbladder stones. I think the findings are very similar to those found on oral cholecystogram. There is still about a 5 percent false negative rate in that if the gallbladder contracts well, there may still be disease present. We do use this occasionally. However, we are relying less and less on the size of the gallbladder because of its extreme variability and really haven't found this particularly helpful.

LEOPOLD: In addition to cholecystokinin, there is an agent that is now available commercially which is a terminal octapeptiate of cholecystokinin called Syncacolide, another commercial name for it is Kinovac, which has its maximum effect if given intravenously. Maximum contraction of the gallbladder occurs approximately 10 minutes after intravenous injection. This brings up the possibility that at least one could give an injection of Syncacolide and watch the gallbladder with real-time and come up with studies that are very much like the ones seen on cholecystokinin, cholecystography. I think it is probably worth looking at. If the data turn out to be anything like cholecystography with cholecystokinin, then we will have a lot more things to argue about.

VERDON: I have another question. In a jaundiced patient if ultrasound cannot demonstrate dilated intrahepatic ducts, how feasible is transhepatic cholangiography?

SAMPLE: The data on this are as follows: You take all the series that have been reported in the literature so far in a nondilated intrahepatic biliary system; people are running about 65 to 70 percent successful entry into the system. Some of the more recent series are in the 90 percent entrance rate. So we may have to wait awhile to see what this shows. The data on ERCP are slightly higher in the nondilated biliary system than all of the data on transhepatic cholangiography. It is in the neighborhood of 80 to 85 percent, whereas in the dilated system, the entrance is less than that, somewhere in the high 50th or 60th percentile. So in our hands, as I indicated in my flow chart, it is of important triage information to find out if the system is or is not dilated. What we do in that case, again depending on clinical information, is usually a liver biopsy. If there are indications on the liver biopsy or there is clinical course or chemical course that still suggests there is a surgical cause or potentially surgical cause to the jaundice, then we have tended to go to an ERCP with a nondilated system. In our own hand, in our own series, our entrance into the liver intrahepatic biliary system is about 80 percent for nondilated biliary and 100 percent for dilated biliary tracts.

VERDON: Are any of you having your patients fast for a regular abdominal ultrasound scan?

SAMPLE: We have all our patients fast for an abdominal ultrasound because of the surprising number of times you come up with biliary disease findings. I don't know that it has helped the gas problem. I think we still have the same percentage of patients that are gassy versus nongassy, but it has allowed us to equivocate far less often about the gallbladder. I agree with Dr. Leopold that we are finding an incredible instance of gallstones in asymptomatic patients. Another factor that I didn't mention but Dr. Leopold alluded to in our series of jaundice patients is that around 25 percent of patients with a medical cause to their jaundice had gallstones that were totally unrelated to their problem. So these cases may be difficult, and the main value of the ultrasound will be that we can make a more definitive statement about the biliary system.

VERDON: When patients come in with suspected acute cholecystitis and you do an ultrasound and don't demonstrate any distinct stones, do you have any other criteria in these patients to make the diagnosis of cholecystitis in the absence of stones?

SAMPLE: We use a very similar criteria to what Dr. Leopold demonstrated. In the patient without ascites if we see any thickness to the gallbladder wall, we feel that this is indicative of cholecystitis or gallbladder disease. The picture of florid acute cholecystitis—here we find a very classic image on ultrasound with a very thickened wall and a very hazy outline to the gallbladder. We have had several examples of evidence of rupture of the gallbladder and a development of a perigallbladder abscess. These are classic findings and again, depending upon the philosophy of the surgeon and whether you demonstrate that it's actually ruptured or not, these patients will or will not go to surgery. Also, we put extreme or complete confidence right now in the absence of visualization of the gallbladder if the patient has any sizable time to fast prior to the time we study them. We find that this is always associated with gallbladder disease.

VERDON How often do you see acute cholecystitis without stones? Do you think this is more common than we realize or is it relatively rare?

LEOPOLD: No, I don't think it is that common. I think more often we will see stones associated with gallbladder disease.

Echography and Other Diagnostic Methods in Retroperitoneal Node Enlargement and Other Masses

Atis K. Freimanis

Masses arising in the pancreas and the kidneys will be excluded from consideration here since they are discussed elsewhere in this volume. I will restrict myself to masses arising from other retroperitoneal structures and especially lymph nodes.

The question of lymph node enlargement arises most commonly in patients with lymphoma, although the need to identify enlarged lymph nodes in addition to other masses exists in other neoplastic diseases as well.

There are four major purposes for imaging retroperitoneal lymph node enlargements:

1. *Detection* in the patient who is being worked up for symptoms and signs of disease either in the retroperitoneum or elsewhere.
2. *Diagnosis* and differential diagnosis.
3. *Staging.*
4. Follow-up and rechecking in the course of *management.*

We will discuss some of the procedures involved as well as their indications, their preferred sequence, relative value, and other considerations.

In addition, we will discuss the imaging diagnosis and differential diagnosis of other retroperitoneal masses. The purposes of study of these masses are similar to those listed above.

Detection of Enlarged Lymph Nodes

Echography

Active ultrasound laboratories in today's practice are examining quite a few patients for the chief complaint of abdominal pain. The pain often is not or cannot be specified more exactly and it is clear on review that one is simply searching for a possible lesion *without* specific indications of a strongly suspected location. In this respect, *each* of the several imaging methods has only a slight chance of showing the lesion. Even the combination of all the available methods leaves one without a specific diagnosis much of the time.

This is most apparent in respect to the plain roentgen film of the abdomen, a regularly accepted and widely used procedure. In the search for the causes of abdominal pain, this examination very often has normal results.

The *searching* or survey type of echographic examination should be performed in a systematic fashion, obtaining cross and sagittal sections at a spacing of 2 cm, extending the examination as high in the abdomen as the interference of the lung permits and as low into the pelvis as the public symphysis. Sagittal sections should also be obtained to each side of the midline as far laterally as satisfactory images can be obtained. At the present, we scan with a 3.5 MHz medium internally focused transducer but drop back to 2.25 MHz in patients in whom penetration is insufficient at 3.5 MHz. The method and marking of the sections are as described by the authors[2] and more recently by the Standards Committee of the American Institute for Ultrasound in Medicine. Lesions explaining abdominal pain may under these circumstances be found in various locations. Unexpected lymph node enlargement may be found in the periaortic area. In women, pelvic masses at times account for unexplained and atypical abdominal pain. In the upper abdominal area, involving the pancreas and the structures surrounding it, the gallbladder and the whole area between the two kidneys, the liver, the spleen, and the gallbladder are common sites for unexpected lesions. Our procedure is to perform the routine study and then to make an attempt to make sure that the areas of particular frequency of disease, as just mentioned, are covered with satisfactory quality, more closely spaced (if necessary) scans. Specific methods for the various organs such as the liver, gallbladder, pancreas, adrenals, and others have been described elsewhere.

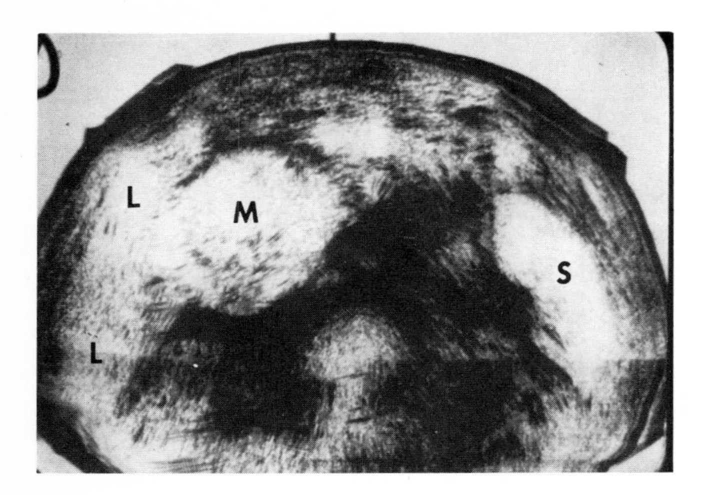

Figure 1. Cross section 4 cm above level of iliac crest. A quite transonic but internal echo containing solid subhepatic mass (M). The liver (L) and spleen (S) are shown as cross sections of their lower edges. A small node may be adjacent to the larger mass in the midline. The excellent sound transmission to the right anterolateral aspect of the vertebral body and the adjacent connective tissues through the obviously solid mass is remarkable and rather characteristic of lymphoma. Diagnosis–lymphocytic lymphoma.

In case *lymphoma* nodes are present, these can appear in certain rather classical manifestations, which include: (a) a dominant, single mass appearance (Fig.1), (b) a preaortic plate or vessel enveloping pattern (Fig. 2a, b, c), (c) bilateral symmetrical periaortic masses (Fig. 3), and (d) multiple spindle-shaped separate masses (Fig. 4a,b,c).[2,6,7]

For identification of these lesions, certain additional features have been observed to be helpful. The presence of an enlarged spleen is in favor of lymphoma as contrasted to other masses. Nodes that involve the root of the mesentery—and we have found these to be fairly frequent in our series of lymphoma patients—characteristically will show what appear to be oval-shaped masses with plates of tissue within these masses (Fig. 4b,c). The plate represents the mesentery or portions of the mesentery. Frequently vessels can be recognized within this plate.

Originally, we found it difficult to identify enlarged lymph nodes in the lesser pelvis. There are still problems in this area, although with improved equipment and better recognition of the anatomical structures, we are able to identify lymph nodes along the iliac chains and elsewhere in the pelvis more reliably than heretofore (Fig. 5a,b).

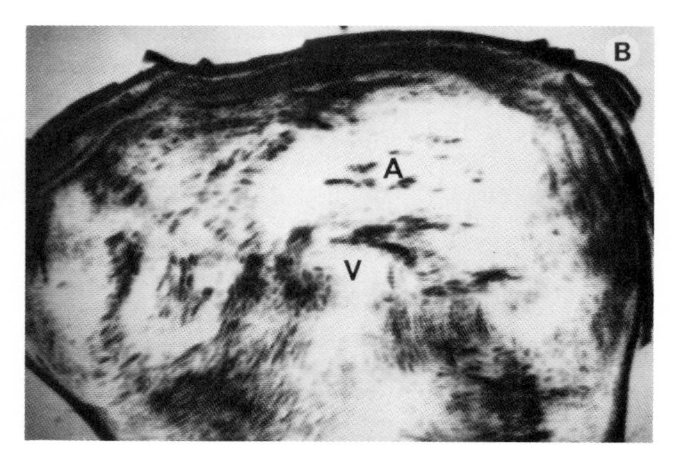

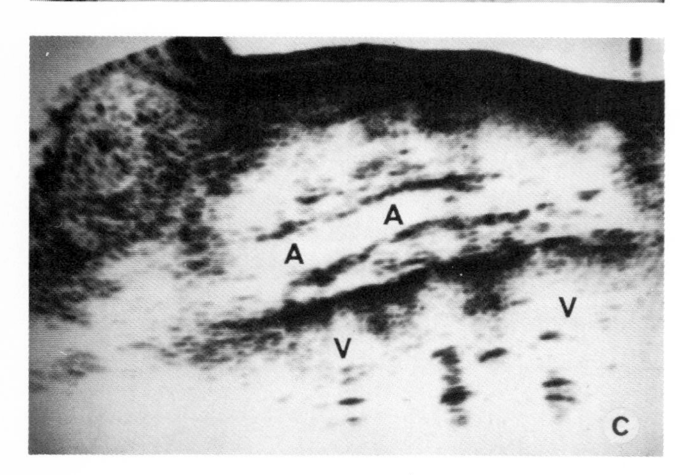

Figure 2. A. Cross section 2 cm above the level of the iliac crest. Hodgkin's lymphoma with a mass (M) surrounding the abdominal aorta (A) and partially obliterating its contour. (V = vertebral body.) **B.** Cross section 6 cm above iliac crest. The abdominal aorta (A) is surrounded by a lymphoma mass both from the front and from the back. It is elevated off the vertebral column (V). Most cases of periaortic lymph nodes do not show this pattern. When present, this pattern is very characteristic of lymphoma, but is also seen with testicular tumor and other lymph node metastases. **C.** Sagittal section 2 cm to left of midline in the same patient.

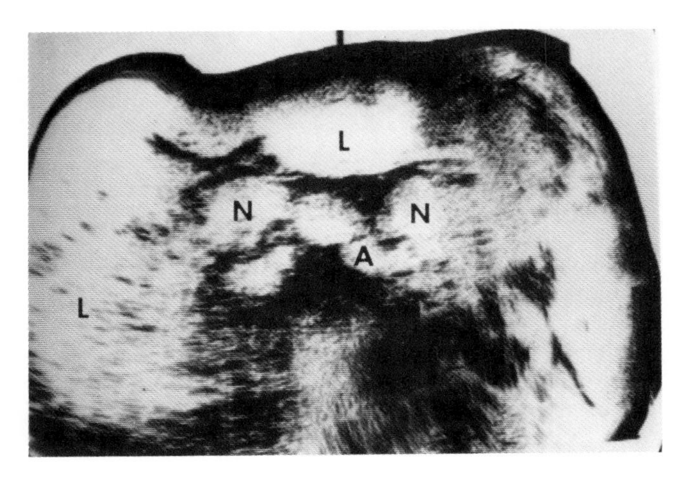

Figure 3. Cross section 9 cm above the level of the iliac crest. Symmetrically enlarged lymph nodes (N) to each side of the abdominal aorta (A). An additional node is interposed between the two. The structure behind the large node on the right may be the inferior vena cava, but one cannot be sure from this section alone. The transonic roundish structure anteriorly in the middle is a somewhat enlarged left lobe of the liver (L).

Reliability of enlarged lymph node detection was estimated originally at 90 percent by us[2] and since has been confirmed by other authors.[4,6]

Computed Tomography

Computed tomography is capable of showing enlarged lymph nodes. It appears to work better in more obese patients because the nodes (as other retroperitoneal organs) are surrounded by more fat and therefore are more apparent. As in the demonstration of the pancreas, there appears to be an advantage for computed tomography in the obese patient and an advantage for ultrasound in the thin patient. Gastrointestinal gas causes only little difficulty for computed tomography, but in some cases it makes echography difficult. While the general principles of computed tomographic scanning are becoming established, a standardized system such as exists for ultrasound is not available at the present time. Also, at the present time sagittal sections are not available. The radiation dose is of some concern in a survey type study since presumably most patients examined in this manner will not turn out to have malignant disease. Thus, even though the method is

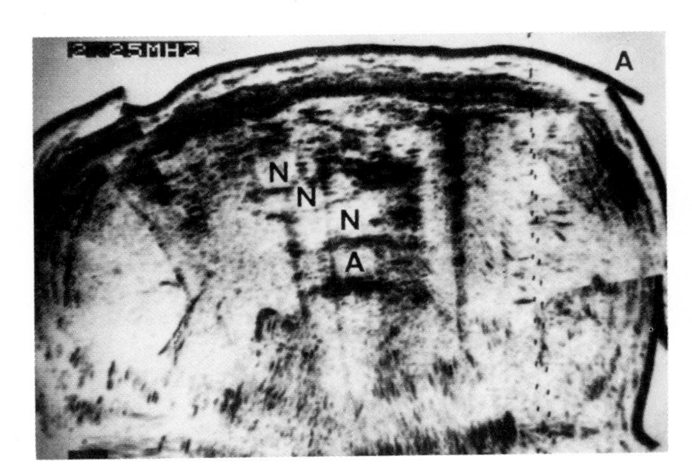

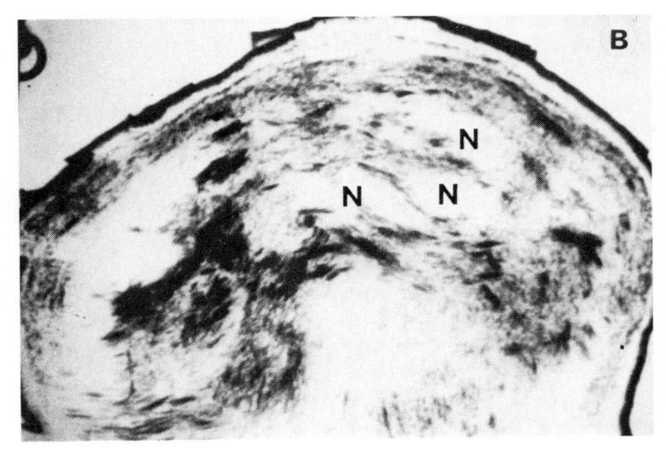

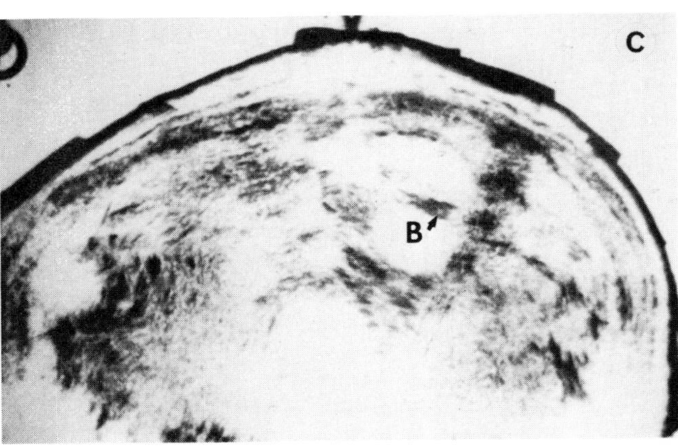

Figure 4. A. Cross section 1 cm above the level of the iliac crest. This is an example of rather minimally enlarged lymph nodes. If such nodes are isolated, they may be hard to identify. In this case there is a whole group so that no other explanation fits. (N, enlarged lymph nodes; A, abdominal aorta.) **B.** Cross section 6 cm above the level of the iliac crest. Elongated, and in cross section somewhat spindle-shaped appearing enlarged lymph nodes (N). **C.** Cross section 4 cm above the level of the iliac crest. Same patient as above. Somewhat spindle-shaped lymph node masses with a portion of the mesenteric plate between them containing a small blood vessel (B).

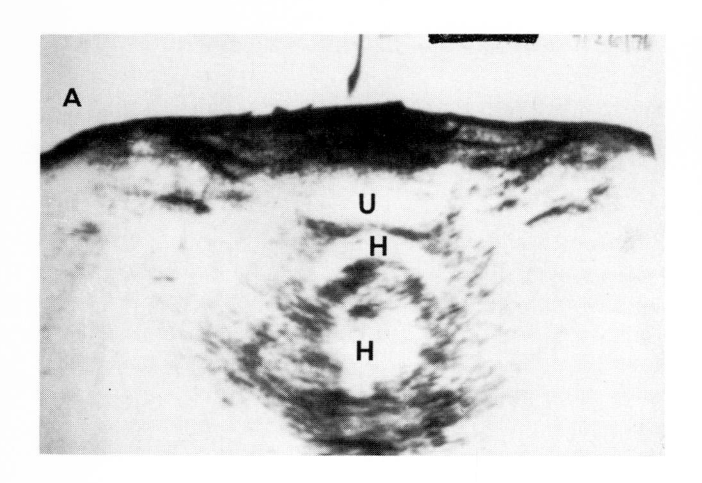

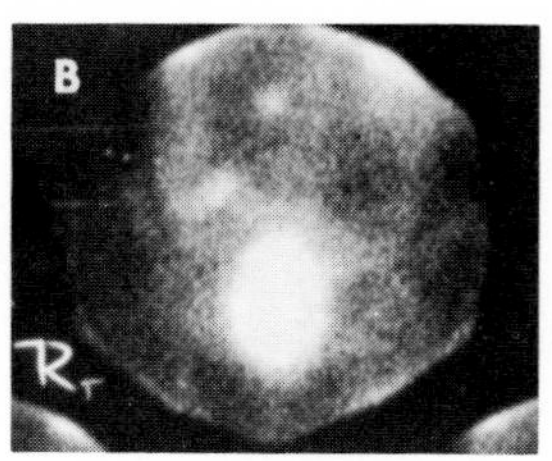

Figure 5. A. Cross section of the lower lesser pelvis in a patient with known Hodgkin's disease. (H, rectal and perirectal Hodgkin's disease infiltration; U, urinary bladder.) **B.** Radiogallium scan encompassing the mid- and lower abdomen. The very active area in the lower portion in the middle of the picture is the uptake of radioisotope by the rectal and perirectal lymphoma infiltration shown on Figure 5A. Note that there is an additional node probably in the paraaortic area on the right and that there is radioisotope uptake in the area of the cecum. This eventually turned out to be additional lymphoma infiltration in that organ.

highly efficacious in finding lymph node enlargements, its place in looking for such enlargement is not totally clear at this time.

Radiogallium Scanning

In contrast to echography and computerized tomography, both of which are tomographic procedures, radiogallium scanning is a method of overall one-view survey imaging. Radiogallium scanning is nonspecific. Increased uptake may be found in certain inflammatory lesions as well as in neoplastic lesions. Many neoplastic lesions do not show increased uptake of the isotope and the uptake is unreliable even in lymphomas. Thus, the test is not reliable in all individuals and may not be reliable for all lesions in a given individual. On the other hand, since the agent is injected systemically, the radioisotope may be picked up in lesions in places where neither ultrasonic nor computerized tomographic scanning is regularly performed. Such locations may particularly include portions of the intestinal tract where wall involvement may be difficult to detect by either ultrasound or computerized tomography, and the various corners and recesses in the abdominal area as well as lymph node locations outside the body cavities. Due to the tomographic nature of both ultrasonic and computerized tomographic examinations, it is and will most likely remain difficult to examine all possible corners of a given patient by these methods (Fig. 5a,b).

Thus, radiogallium examinations provide an overall, if somewhat unreliable, searching tool. In fact, Braschco[4] states that the combination of ultrasound and radiogallium scanning is one of the most efficient ways of detecting lymph node enlargements. The same presumably would apply to a combination of radiogallium scanning and computed x-ray tomography if the latter could be performed in an extensive fashion. The major drawback of radiogallium scanning is the absence of visualization of diseased nodes in a certain percentage of patients.[10,12]

Lastly, it may be pointed out that radiogallium examinations do have an additional advantage—they may in some cases identify a lesion as being present regardless of whether it is neoplastic or inflammatory. This, in certain circumstances, is helpful in identifying the cause of unexplained fever or other clinical symptoms if the lesion is not readily found by the location of the symptoms and other methods.

Sequence of Examinations

What sequence of examinations should one perform? The conventional arrangement has been to obtain plain films of the abdomen, cholecystograms, intravenous urograms, and barium studies first and then proceed to the more complex, sometimes invasive examinations such as radioisotope studies, ultrasound, arteriography, and others.

It appears that it may now be quite appropriate to change the sequence of these examinations. Plain roentgenograms are still very valuable and should be obtained early in the course of events. The same applies to the use of soluble iodine containing contrast media—such as the oral cholecystogram or the intravenous urogram. If satisfactory excretion and concentration occurs, these methods are still among the best and most exact available.

In respect to the gallbladder, it needs to be noted that ultrasound may be very helpful and it is now justifiably recommended as the next best test after a single nonvisualized cholecystogram. We quite agree with this recommendation on the basis of our experience. It is further being recommended, and we believe that we will see a shift in this direction, that an ultrasound cholecystogram be the *first* test for suspected gallstones. Once the echogram is being performed for gallstones, a search for *retroperitoneal and other lesions* may well be performed at the same time. We strongly recommend and in our laboratory perform survey scanning in most cases, even when the consultation request is primarily directed at a single organ examination.

In any case, it is our recommendation that in addition to the plain films and excretory iodine studies, ultrasonic examination be performed as one of the *first* tests in patients presenting with abdominal diagnostic problems.

The place of CT scanning in the sequence is not quite clear. The test is presently so expensive that one hesitates to recommend it as one of the first procedures. Doing it first would obviously lead to a great number of these tests being performed. It may well be that in selected cases eventually a place will develop for computerized tomography somewhere in conjunction with the other tests which involve injection of contrast media—such as the intravenous urogram—because of the benefits of enhancement in many organs.[8]

We believe, on the basis of our experience, that in a smaller, primary and secondary care hospital, ultrasound will have a *low* frequency of detection of unsuspected retroperitoneal masses, whereas the detection of gallstones and renal cysts will be high. In a tertiary care, university, or large city hospital, ultrasonic examination will be helpful in detecting unsuspected and otherwise not demonstrated retroperitoneal masses with a frequency sufficient to make echography a desirable examination early in the sequence of studies in diagnostic problems in the abdomen.

Somewhere, one will engage in a dilemma of whether it is, in a given patient, better to attempt examinations which are either targeted at a specific suspected diagnosis or which are expected to be most helpful in a given case, as contrasted to the rapid performance of multiple screening tests. The second approach has the disadvantage of high total cost for the examinations. On the other hand, if the various tests are done rapidly, the cost savings in hospitalization may be considerable and easily may outweigh the targeted examination procedure; the earlier establishment of a diagnosis is an associated, nonfinancial consideration. To what extent each additional test adds to the diagnosis, remains questionable. It may vary with the disease the patient has.[3,4,9,11]

An additional concern in deciding this last question is the possible harm arising from the various examinations. One should remember that an intravenous urogram is presumably not harmful except for the slight radiation exposure, but it may on occasion—very rarely—be associated with a severe contrast media reaction which may be fatal. Fatal reactions are estimated to occur once in approximately every 40,000 intravenous urograms. Computed tomography is associated with a radiation dose in the general range of diagnostic roentgenology, but distinctly on the high side. It gives less radiation exposure than arteriography. There is no immediate danger to life from the procedure itself. The incidence of severe reactions is likely to be about the same as with intravenous urography, if enhancement with contrast media is used. Contrast agent reactions are not a problem with radiogallium scanning. It is associated with moderate radiation exposure.

Staging

The diagnostic ability of both ultrasound examinations and computed tomography in lymphadenopathy at the present time is size related. That is, normal size lymph nodes with somewhat abnormal internal architecture will not be readily picked up by either method. Therefore, in primary staging, these methods are not sufficient by themselves. Additional studies, including lymphography and possibly laparotomy, are necessary to complete the staging procedures. On the other hand, both by ultrasonography and by computed tomography, lymph node masses in areas not filled by lower extremity lymphography may be identified. These areas include particularly the subhepatic area, lymph nodes high in the left upper abdomen in the general area between the spine and stomach, and, on quite a few occasions, the root of the mesentery. Some of these areas are more accessible by computed tomography, some by ultrasound.

Our present recommendation for radiological studies in a patient with known lymphoma for *staging* of abdominal disease includes intravenous urography, lymphography, barium gastroduodenal examination, and echography. Laparotomy is done when necessary.

Diagnostic Studies in Management

While lymphography is quite valuable in following those lesions which are visualized by it and while it shows internal patterns, in the follow-up management of lymphoma the geographically more extensively capable searching tools have advantages. If disease recurs, it does so more often outside the periaortic chain in the course of the later stages of the disease than at first.

Thus, echography, radiogallium scanning, and computed tomography have the advantage of showing disease in areas not visualized by lower extremity lymphography. In our hands, echography has been most useful in following of lymph node enlargements once they are identified. The echogram at the time of first staging makes a very useful baseline. Examinations can be repeated at will with no pain, no intervention, and at reasonable cost in both hospital and ambulatory patients. If the location of the mass is *known* the follow-up becomes quite easy. It does need to be pointed out that we do a full survey examination of the abdomen on each recheck. We do observe at times that unexpected nodes appear in areas outside the area of direct interest being followed at any particular time. *Discrepant behavior* of various masses can occur with various neoplastic diseases. This is not so much characteristic of lymphoma (although it does occur in this disease occasionally); with lymph node enlargement or extranodal metastases due to other neoplasms it is not at all uncommon to see some masses enlarging while others are decreasing. Intravenous urograms, gastrointestinal series, and other studies are still obtained and have a place. Barium studies, however, are most efficiently performed toward the end of the sequence of diagnostic examinations. Many of these diseases eventually start involving the intestinal tract as well, and barium studies are very helpful in the evaluation of these. However, with the advent of lymphography, echography, and gallium scanning, the frequency of intravenous urograms and gastrointestinal series in the management of lymphoma and metastatic tumor has decreased.

This at first was not readily apparent to us. However, observation of our recent practice indicates that in quite a few of our patients we find ourselves simply doing echograms at regular intervals and not much other radiographic study unless circumstances dictate otherwise.

The value of CT scanning in management is not quite clear. Obviously, the test is very capable. At the present time, it appears most valuable in following specific known lesions, especially if these turn out to be difficult to show by other methods.

We believe that the practice of medicine has a strong self-regulating and self-adjusting evaluation mechanism for diagnostic and therapeutic procedures which is not readily apparent to the casual or external observer. We also believe that this process is rather slow—much slower than planned efficacy research—but ultimately much more reliable and beneficial. This is *not* to say there is no place for designed research. One should not, however, overlook naturally existing processes. Many of our "planners" have turned out to be greviously wrong.

Imaging Tests for Diagnosis and Differential Diagnosis

This section will deal more with differential diagnosis than with detection. While it is still true that it may be difficult to tell one mass from another on any one examination—such as computed tomography or ultrasound, accumulating experience in the combination of methods suggests that the *divergence of findings* on these examinations will increasingly contribute in a major manner to the differential diagnosis of retroperitoneal lesions. This concept has now been described in terms of the ultrasound examination being helpful in separating isodense (by CT) masses from each other.[3,5] Conversely, in certain circumstances the computed tomographic x-ray examinations will separate isoacoustic masses from each other. Additional contributing procedures such as intravenous urograms (excretory function and displacements), gastrointestinal series (luminal contour and mass effects), and isotope studies (chemical function and composition) yield further differential diagnostic possibilities.

Our observations clearly indicate that while logically many of these differential diagnostic capabilities exist as soon as these various methods are introduced, it takes time for people in the field to develop a feel for the value and the relationship of these procedures and to develop an ability to analyze the results. I do rather strongly believe that we are seeing a distinct change in our ability to analyze epigastric deep abdominal and retroperitoneal lesions by a *combination* of these methods. It appears that the most valuable diagnostic results are obtained in those laboratories where computed tomography, echography, and radioisotope studies are located in reasonable geographic proximity so that the results can be compared readily.

We are, in fact, in an era in which our younger generation of physicians is growing up with an assurance that visualization—by imaging procedures—of the pancreas, the liver, the celiac and mesenteric vessels, the lymph nodes, and other deep intraabdominal structures is—to them—a naturally expected, basic part of medical practice. Instead of trying to suppress and limit the availability of these tests, we should be working on making these tests cheaper, more readily available, and more interactive. Progress cannot be stopped. It is desirable to direct it so that the patient and society benefit most.

Angiography

This discussion has not centered much on angiography. Angiography is an invasive method. Used by capable hands, it has proven to be a *major* contributor to the diagnostic process and is reasonably safe. We are now convinced that many lesions in the upper abdomen can be diagnosed and treated effectively without exploratory laparotomy. Our observations in this area include lesions such as cysts of various kinds, abscesses, and benign and malignant neoplasms, which can be biopsied either directly or in metastatic implants. Even in those cases where surgery is carried out for diagnosis and possibly for treatment, the accuracy of location and nature of the expected lesion as provided to the surgeon has shown tremendous improvement. The ability to diagnose periaortic masses as contrasted to an aneurysm or to other lesions which cause ureteral obstruction, as examples, has been very helpful. Other interesting examples include the various adrenal tumors, metastatic lesions in unusual locations in the abdomen, and benign tumors arising in the retroperitoneal area.

Other Retroperitoneal Masses

Primary neoplastic lesions in the retroperitoneal area are infrequent. Nevertheless, when they occur, they may be difficult to detect, localize, stage, and differentially diagnose. Metastatic lesions also occur in this area. While they are much less common than lymph node metastases discussed in the previous sections of this chapter, occasionally they do occur and again are frequently difficult to identify by conventional means.

In our experience, echography has been most helpful in evaluation of lesions in this category. Computed tomography also most likely will turn out to be quite helpful. Since most of these organs and lesions are "water density" tissues, standard radiographic procedures often are of little value.

Retroperitoneal abscesses of various kinds, including psoas abscess, retroperitoneal hemorrhages, and retroperitoneal tumors, both benign and malignant, can be identified and outlined. A histological diagnosis is not readily possible because the internal acoustic characteristics of the lesions vary. However, one is generally able to separate the abscesses, urinomas, and hematomas from the solid lesions echographically.

Identification of the lesion and its blood supply, both of which may be important for surgical treatment, frequently require angiography which is often very helpful in this part of the body (Fig. 6a,b).

The scanning procedure in our echographic practice is the same as for abdominal lesions. For one thing, retroperitoneal structures are to a great extent seen well from the front. Also, it is worthwhile scanning the abdomen from the front to exclude intraabdominal lesions. We also scan these patients from the flanks or from the back as may be appropriate. In that case the scanning procedure, spacing, and slice identifications are similar to those from the front.

A most helpful item in our daily ultrasound practice has been the development of a sheet with a set of body diagrams on which the direction of any unusual scan planes can be indicated and the corresponding images so identified. We have found this to be much more efficient for the various angled and inclined planes than numerical or verbal descriptions.

Conclusion

Echography, radionuclide scanning, computed tomography, and arteriography altogether are changing our view of medicine. Areas previously considered unaccessible to diagnostic imaging can now be imaged with considerable detail. The current fuss about relative efficacy of various procedures to some extent will turn out to be pointless in that the combination of several imaging modalities does not simply show the same thing better more times but often produces much more accurate diagnostic information by virtue of diagnostic application of the discrepancy of findings. The most efficacious procedure for the patient's benefit would be a rapid, logical, and coordinated application of these various methods. A prompt diagnosis is one of the most

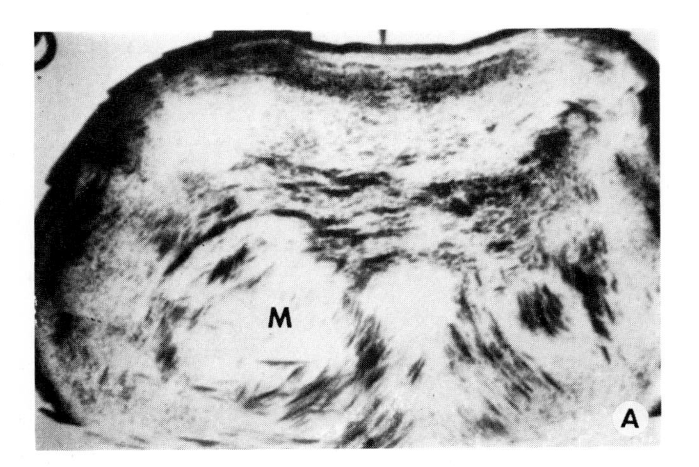

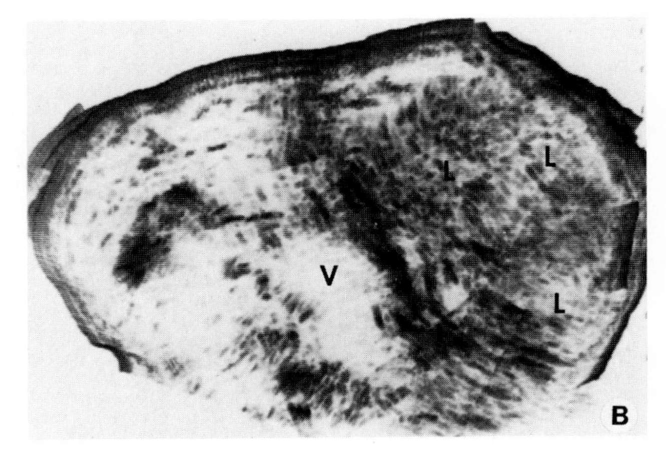

Figure 6. A. Cross section 10 cm above iliac crest in patient with known melanoma. Retrorenal metastasis on right (M). The right kidney is displaced anteriorly. **B.** Cross section 5 cm above iliac crest in a child. Large surgically proven benign lipoma (L) occupying most of the left side of the body. (V, vertebral body.)

efficient ways to contribute to the patient's welfare and economy of medical care.

References

1. Alcorn FS, Mategrano VC, Petasnick JP, Clark JW: Contributions of Computed Tomography in the Staging and Management of Malignant Lymphoma, Radiology 125: 717, 1977
2. Asher WM, Freimanis AK: Echographic Diagnosis of Retroperitoneal Lymph Node Enlargement, Am J Roentgenol 105: 438, 1969
3. Birnholz JC: On Maps and Comparing Cross-Sectional Imaging Methods, Am J Roetgenol 129: 1133, 1977
4. Brascho DJ, Durant JR, Green LE: The Accuracy of Retroperitoneal Ultrasonography in Hodgkin's Disease and Non-Hodgkin's Lymphoma, Radiology 125: 485, 1977
5. Bryan PJ, Dinn WM: Isodense Masses on CT: Differentiation by Gray Scale Ultrasonography Am J Roentgenol 129: 989, 1977
6. Filly RA, Marglin S, Castellino RA: The Ultrasonographic Spectrum of Abdominal and Pelvic Hodgkin's Disease and Non-Hodgkin's Lymphoma, Cancer 38: 2143, 1976
7. Freimanis AK: Echographic Diagnosis of Lesions of the Abdominal Aorta and Lymph Nodes, Radiol Clin North Am 13: 557, 1975
8. Goldberg HI, Korobkin MT: CT, Echo before PTC, ERC, Gastroenterology 72: 190, 1977
9. Gregg EC, Rao PS, Friedell HL: An Analysis of the Value of Additional Diagnostic Procedures, Invest Radiol 11: 249, 1976
10. Hayes RL: The Tissue Distribution of Gallium Radionuclides, J Nucl Med 18: 740, 1977
11. Hillman BJ, Hessel SJ, Swenssen RG, Herman PG: Improving Diagnostic Accuracy: A Comparison of Interactive and Delphi Consultations, Invest Radiol 12: 112, 1976
12. Johnston GS, Go MF, Benua RS, et al.: Gallium-67 Citrate Imaging in Hodgkin's Disease: Final Report of Cooperative Group, J Nucl Med 18: 692, 1977

Radionuclide Imaging, Computed Tomography, and Gray-Scale Ultrasonography of the Urinary Tract

Zachary D. Grossman
Brian W. Wistow
Patrick J. Bryan
William M. Cohen
Frank Seidelmann

While computed tomography, radionuclide imaging, and ultrasound compete for similar information in the study of hepatic space-occupying disease, they are complementary to a much greater extent in study of the urinary tract.

In the majority of distal urinary tract diseases and anomalies, standard radiographic techniques (intravenous urography, voiding cystourethrography, and retrograde pyelography) demonstrate both the functional and anatomic problems underlying urinary tract incontinence or obstruction. The three newer modalities offer little in such evaluation. However, the standard radiographic techniques usually cannot determine the cause of extrinsic bladder compression, while both CT and ultrasound often clearly identify the offending normal or abnormal organ (Fig. 1). With the full bladder as a "window," ultrasound is particularly able to distinguish the uterus from other pelvic masses, whereas delineation of the uterus apart from adjacent pelvic structures is often difficult with CT. This problem is diminished by insertion of a vaginal tampon prior to the examination, identifying the upper vagina and its relationship to other pelvic organs (Fig. 2).[1]

Radionuclide imaging of the distal urinary tract is

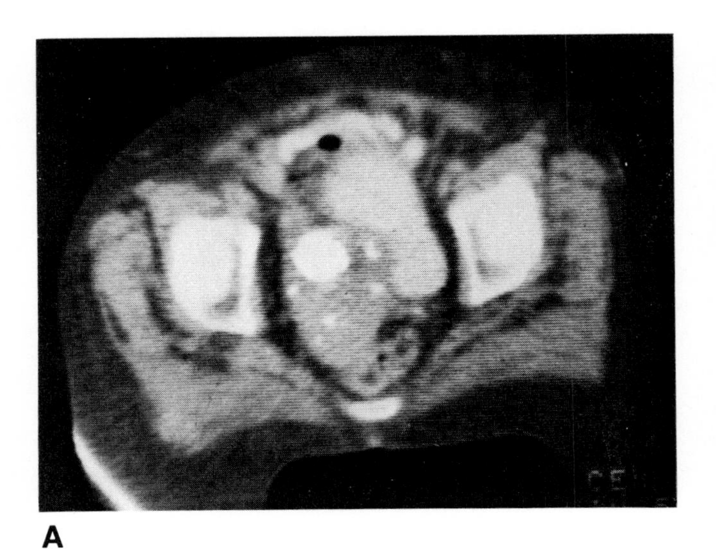

A

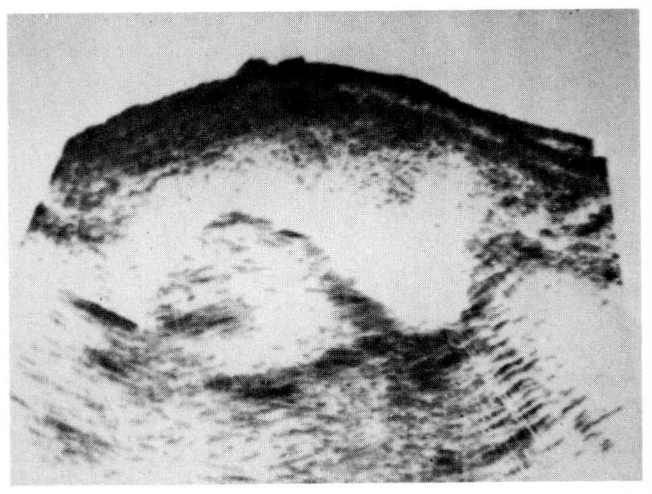

B

Figure 1. **A.** CT of pelvis demonstrating indentation of the urinary bladder from the posterolateral aspect by a large uterus with multiple fibroids. **B.** Transverse sonogram demonstrating urinary bladder indented posteriorly by a large ovarian cyst.

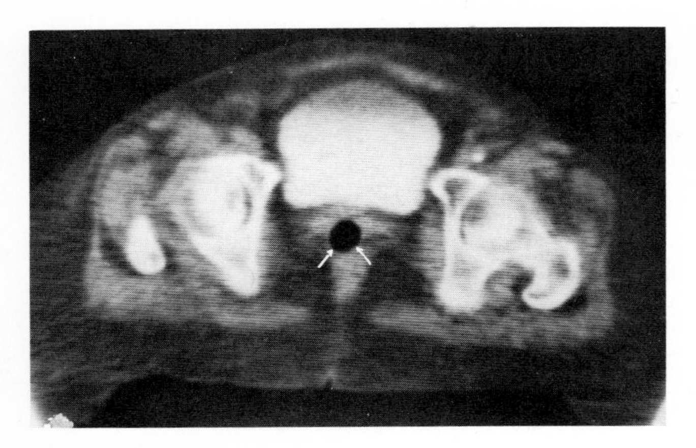

Figure 2. CT of pelvis at the level of the femoral heads with iodinated contrast material present in the bladder. The tampon present in the superior portion of the vagina clearly delineates the position of that structure in relation to the urinary bladder.

very limited. Postvoid residual urine may be quantitated—though such determinations are rarely requested. Occasionally, the cause of hematuria—e.g., bladder or prostate trauma—may be suggested by bladder distortion or distal ureteral obstruction (Fig. 3).

In the study of the bladder itself, ultrasound is largely limited by instrumentation. The transrectal probe is unlikely to achieve wide patient acceptance. The work of Seidelmann, et al.[2] has established that CT, however, often exquisitely visualizes the bladder wall, perivesical fat, and any excrescences along the mucosa. The superb visualization of bladder neoplasms extending both into the lumen (Fig. 4A) and through the bladder wall (Fig. 4B) depends largely upon installation of CO_2 into the bladder via transurethral catheter, intravenous injection of contrast material, and a variety of imaging manipulations—including supine and prone examination, angulation of the gantry, and bladder evacuation prior to the procedure. CO_2 rather than air is necessary in order to minimize the possibility of air embolism. Simultaneous visualization of pelvic and lower abdominal lymph nodes is also feasible, so that, in addition to staging bladder neoplasms and analyzing pelvic organ displacement or enlargement, CT aids the radiotherapist in treatment planning.

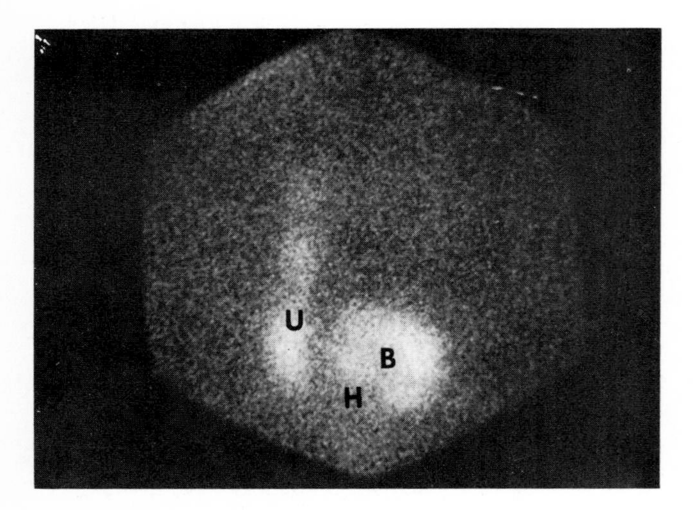

Figure 3. Anterior radionuclide image of the pelvis 20 minutes after intravenous injection of ^{99m}Tc- glucoheptonate. The large indentation of the right side of the bladder represented a hematoma, producing partial distal obstruction of the right ureter (U, uretha; B, bladder; H, hematoma.)

Above the bladder both ultrasound and CT may establish the presence or absence of retroperitoneal adenopathy and clarify apparent ureteral displacement. Because of intestinal gas, ultrasound is considerably less effective in the lower abdomen than CT. Frequently, apparent distal ureteral displacement results from normal variations in the mass and position of retroperitoneal muscle or fat (Fig. 5). While such possibilities may be raised by standard uroradiographic methods, CT is often definitive. More superiorly, in the upper half of the abdomen, both CT and ultrasound clearly reveal retroperitoneal space-occupying disease and its effect upon the urinary tract.

In the kidneys, the radionuclide method evaluates tubular function and perfusion and provides moderate anatomic detail, whereas ultrasound evaluates structure. CT with intravenous contrast injection provides some functional information but is primarily anatomic. Serial transplant examinations, hypertensive studies after a saralasin screening test, and pretransplant donor evaluations represent three areas well covered by scintigraphy. The estimation of renal perfusion (Fig. 6) is particularly critical in transplants, when a life-threatening vascular occlusion may be present and yet undetected by the "anatomic" modalities as opposed to the "functional" modality. Furthermore, since biochemical renal function tests remain normal in the presence of even severe unilateral renal disease, as well as in mild to moderate bilateral renal disease, a variety of conditions warrant repeated monitoring of renal tubular function by the radionuclide method; children followed yearly for known vesicoureteral reflux typify this group. Many urologists, in addition, insist upon evaluation of bilateral function prior to unilateral nephrectomy.

While radionuclide imaging is moderately accurate in determining the vascularity of proven renal masses, the determination of tissue attenuation coefficients by CT and echogenicity by ultrasound have almost eliminated the radionuclide flow study as the follow-up examina-

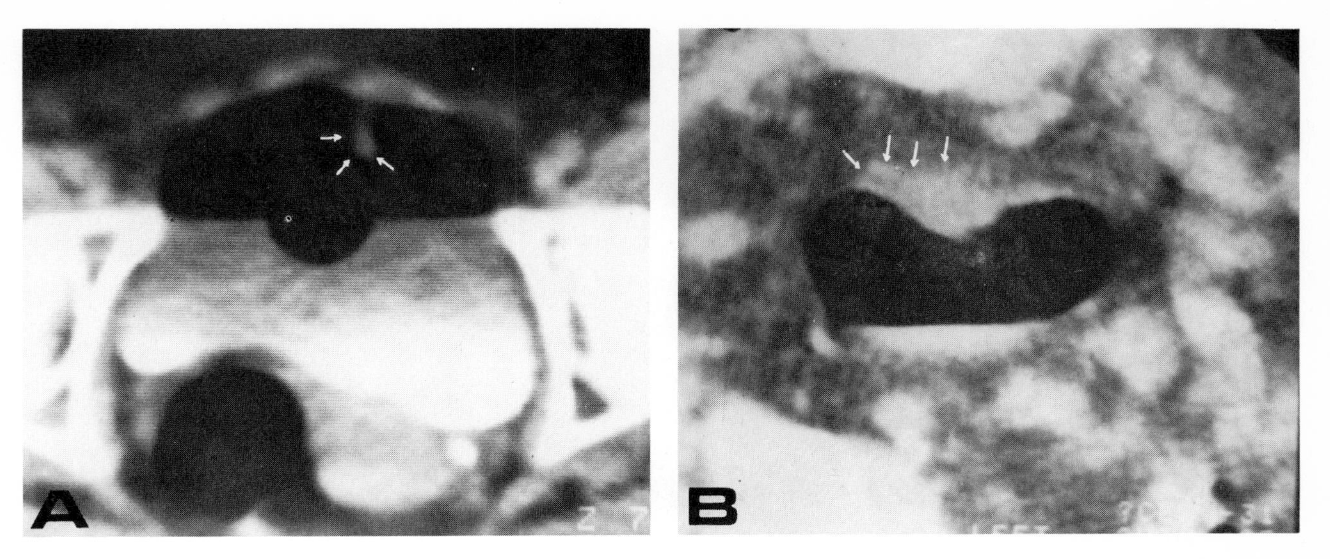

Figure 4. A. Supine CT of the pelvis, enlarged, with contrast material, catheter balloon, and Co_2 in the bladder clearly demonstrates an anterior pedunculated carcinoma of the bladder (arrows) extending into the lumen. **B.** Prone CT demonstrates a bladder wall tumor with irregular extension into the surrounding soft tissues.

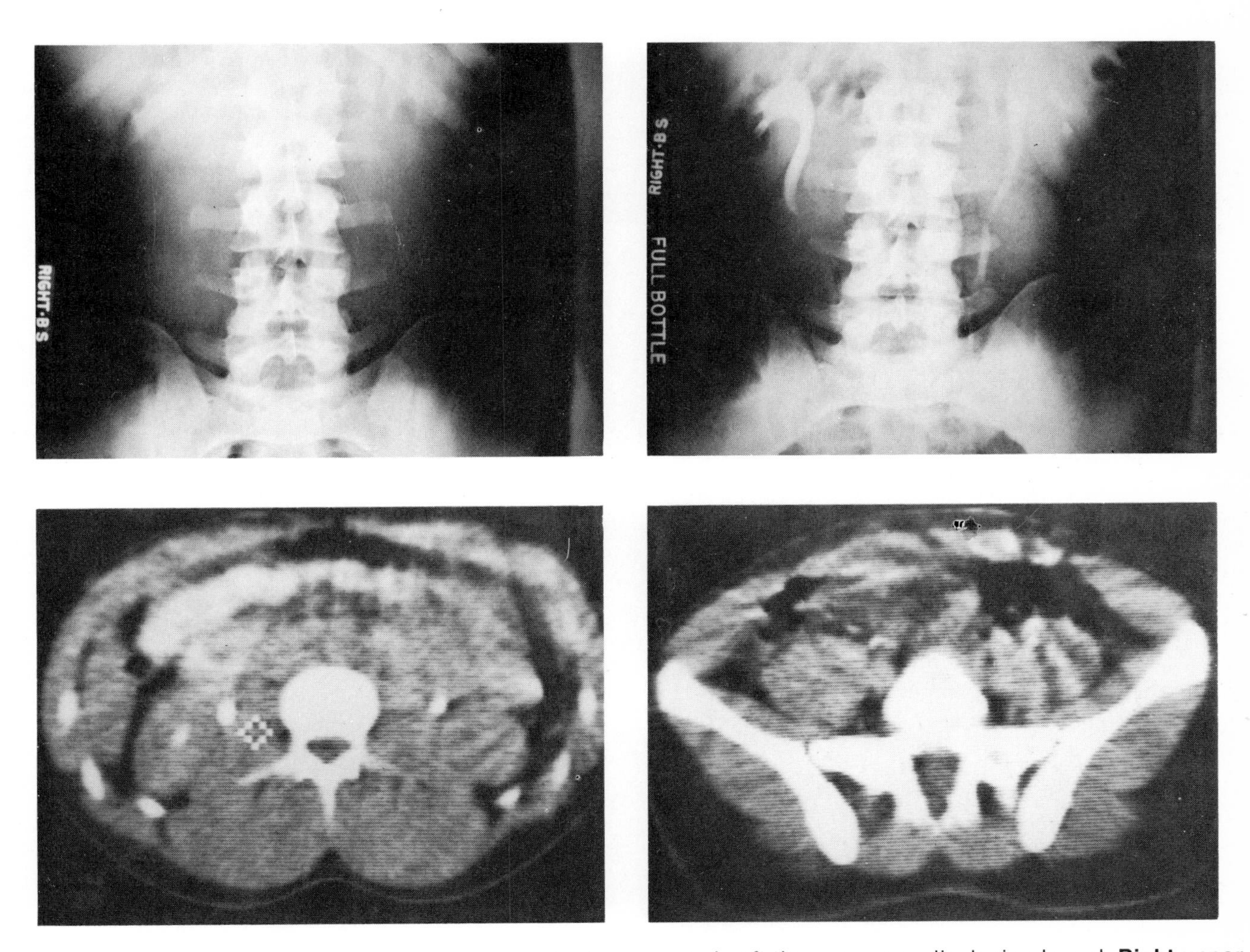

Figure 5. Left upper. Plain film of the abdomen with large central soft tissue masses displacing bowel. **Right upper.** IVP demonstrates right ureteral displacement. **Left lower.** CT of midabdomen reveals grossly enlarged psoas muscles in this young karate-practicing farmer! Displacement of the right ureter and bowel compression by these large psoas muscles is noted. **Right lower.** CT of pelvis again demonstrates very large psoas muscles.

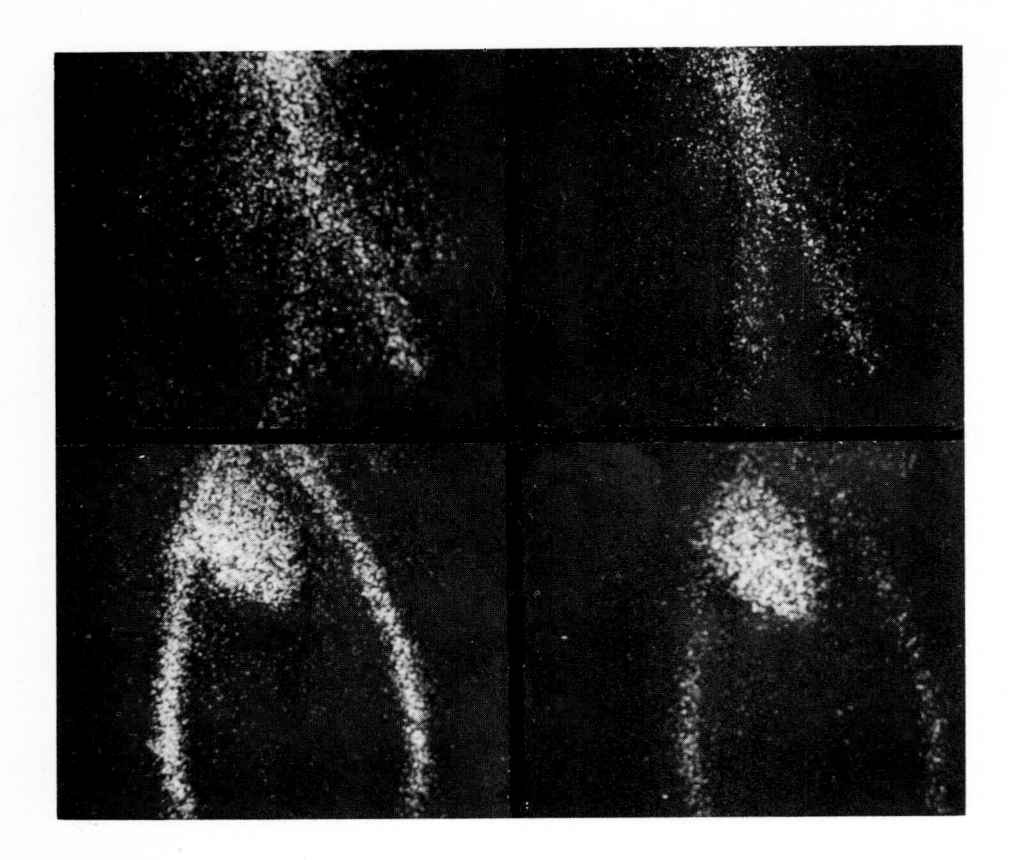

Figure 6. Upper: Radionuclide flow study of the renal transplant in a young male—suddenly oliguric shortly after surgery. No perfusion is observed in the region of the transplant. **Lower:** Radionuclide flow study after the kidney was surgically repositioned to unkink the renal artery, resulting in excellent transplant perfusion.

tion after urographic demonstration of a renal mass. On the other hand, in occasional cases, attention is first directed to the kidneys by accumulation of gallium-67 citrate during a search for inflammatory processes. Subsequent CT and ultrasound further define the underlying abnormality.

Because standard uroradiography is extremely effective in the diagnosis of obstruction, ultrasound and CT are best applied to the question of obstruction in the radiographically nonfunctioning kidney. When technical factors limit such visualization, radionuclide imaging may succeed, and, in addition, the presence or absence of vascular occlusion as an underlying etiology for nonvisualization may be determined.

In conclusion, then, tubular function and perfusion are best evaluated by nuclear imaging; the bladder, pelvis, and low retroperitoneum, by CT; the high retroperitoneum, by CT and/or ultrasound; obstruction, by ultrasound and CT when feasible, and by the radionuclide method if the others fail; and renal space-occupying processes, primarily by CT and ultrasound.

References

1. Cohen WN, Seidelmann FE, Bryan PJ: Use of a Tampon to Enhance Vaginal Localization in Computed Tomography, Am J Roentgenol 128:1064, 1977
2. Seidelmann FE, Bryan PJ, Temes SP, et al.: Computed Tomography of Gas-Filled Bladder—Method of Staging Bladder Neoplasms, Urology IX:337, 1977

Discussion: Renal Imaging

Moderator: Atis K. Freimanis
Panelists: Brian W. Wistow
Hirsch Handmaker

FREIMANIS: How useful is computed tomography in evaluating bony lesions in the abdomen and the pelvis? Would anyone of you gentlemen like to answer this question?

WISTOW: I omitted pointing out on one of the slides that the gentleman with gross metastases also had obvious metastases in his spine picked up by the CT. It is useful to look for them, I think, without a doubt. The problem is that in order to look at the bone, you have to change window settings completely from normal abdominal studies and most people tend to forget to do that. So often we do not look at the bones where we should. There are normal appearances and anatomic variants that could simulate metastatic disease, but I think that these have been moderately well worked out. Identifying dense sclerotic areas in vertebral bodies is definitely an advantage of the CT. I do not have numbers for it but it can be very useful and the bones should be looked at when this type of study is done.

FREIMANIS: Let me add to that. Tomography will certainly show the lesions. It is, however, hard to *search* large areas by any tomographic process. It is recognized that in looking for metastases, the radioisotope bone scan is probably the most sensitive method. Metastases may show up on radionuclide scans before there is much loss or gain of bone substance.

WISTOW: I think that is very true. I would not argue with that at all. It is just that if you are doing the CT, I think one should look for bone lesions also.

FREIMANIS: How about nuclear studies for renal bumps? That is a good question. I suppose these bumps are most commonly found on the left side, but occur also on the right. What do you do about it?

HANDMAKER: The papers on pseudotumors of the kidney gave us a whole new collection of patients to look at. Renal radionuclide scanning is a useful procedure after the ultrasound. Again, I think we have to emphasize what we want to do in most of these patients: (a) Give them what they are paying for, (b) keep radiation exposure at a low level, and (c) not get into a more invasive but unnecessary procedure because you say you do not know. One of the things I did not mention in my talk, and I think maybe this is a good time to mention it, is that if you do a lot of these procedures, you end up forgetting what the question was in the first place. The patient ends up having four times as many procedures. This is sort of the medical "Peter Principle."

FREIMANIS: If one is dealing simply with lobulations, ultrasound may confirm the integrity of the kidney. If it is intact at the edge and the texture is not changing, it is a normal kidney. Excellent hands and mind are needed because the general opinion today is that the ultrasound examinations tell whether the lesion is cystic or solid. Ultrasound renal texture studies are coming. Then we will be able to recognize abnormal renal texture. In today's average case, it still is the problem. Yes, you say, it is a bump and, yes, it is solid. But what is the next step? I would think, at least in our hands, the selective arteriogram is the answer. It is invasive and it is probably the most expensive. But in terms of looking at the renal vascularity of a solid appearing mass, it is the most exact method available.

HANDMAKER: Given improved resolution of isotope scans, I think the point that Dr. Wistow just made is probably true. A renal scan is probably the best way to look at function, to see if the bump functions the same as the rest of the kidney. Unfortunately, resolution is not always good. I think we get into trouble by calling a vague area. I have seen in my own hands a lobulation which I felt represented a scar or an irregular edge. Then we went to an angiogram where we probably could have quit if we had just said: "It looks like a bump, but I am not sure."

WISTOW: Generally, I agree; but one can get a good outline of the renal cortex in those cases, usually showing the shape of the kidney to be very similar to that on the IVP; I think it is useful in that situation to do the nuclear medicine imaging.

FREIMANIS: The question of *when to stop* is a very difficult one. We, as many others, run into the problem of doing a $2,000 work-up in patients only to find nondisease. On the other hand, if one does the IVP, or any other study for that matter, *as a routine procedure to detect early disease*, there is no good way of not following up if an abnormality is found. Then, it turns out to not be confirmed, and one ends up saying: "Well, I really do not think we will do anything about that."

HANDMAKER: No one was really defending the IVP here and Dr. Freimanis, you are probably left in that position. The zero-minute film in our hospital is probably the most important single film done on an IVP. No one really talks about that much anymore because it is out of vogue and a lot of people assume everybody does it. With 30 or 40 milliliters of contrast material and doing a film immediately in that zero minute as a bolus goes through the kidneys, we have seen incredible things in terms of arterial lesions. Often you get a look at both renal arteries, and a brilliant nephrotomogram. For looking at these things, this is a valuable tool without nephrotomography. Just the zero-minute film—you have learned an awful lot. We tend to forget about *basic* things.

FREIMANIS: If you study intravenous urogram films hard enough, you also pick up some opacification of the liver, the spleen, and other organs; and possibly masses. These should not be forgotten. I have the uncomfortable feeling that we are not looking quite as hard at the roentgenogram now that these fancy studies are available. In fact, sometimes it would be cheaper and less painful to the patient if we would look harder the first time—on the first examination. Let me ask a question of Dr. Handmaker: Some years ago we heard a lot about doing renal functional studies with a combination of agents and then sending the data to a computer center near San Francisco, I believe. They would send back a write-up that would tell exactly how the kidney is working. Will you comment?

HANDMAKER: I refuse to defend all of San Francisco! I think there was—there still is a group doing a lot with data processing. I would not give their initials because we do not want to be commercial, but that was a group looking at computer analysis data from hippuran renograms in an inexpensive way. They found a lot of information just from the renogram. They invented a chair and had people lie in this chair. The data were transmitted to a central computer. The problem was, as I see it at least, that they had models for different diseases in their computer and they matched the data from your patient to the models. Assumptions were made from that, and what statistically came closest to their model was said to be in that category. For instance, renal vascular hypertension or renal agenesis. We sent a patient (who, we knew, had only one kidney) for data analysis. The probe looked at the background activity and came back with milliliters per minute of creatinine clearance. It made nice estimates of renal function, but the problem was they were not looking at what they were seeing. They were looking only at where the kidney should be and if there was no kidney there, you got a number anyway. Their system does basically what almost everybody who is doing computer analysis of hippuran clearance does.

The University of California and a lot of people in the East, as well, have been looking at hippuran clearances. It turns out that it is the first 3 minutes that really counts. I think Mike Hayes down here at UCLA has done some beautiful work to show that if you want to look at function in a computerized way, you really only need look at the first 1 to 3 minutes.

People like Dr. Freedman, who are in the audience, probably could comment better on the mathematics of these techniques, if you care to hear them. The answer to the question about the group in San Francisco is, I think, that they were one step short of truth. That is, they were taking mathematical information and trying to make it look like some standard set of patients that they had developed. The error is, they were not really looking at the anatomy of the kidney and flagging it. They were not really looking at cortex and telling you exactly what was going on. They were approximating renal function in milliliters of flow per minute and then extrapolating back.

FREIMANIS: Is Dr. Freedman here? Would you like to comment?

FREEDMAN: I am quite interested in whether or not renal transplants could be better evaluated by computer analysis of the renogram. For about 50 patients, we did serial imaging, and stored the data. We did a variety of mathematical analyses. It became quite clear after a very short period of time that the images were providing all we needed in terms of clinical information and a tremendous amount of time and effort were put into an analysis of these reams of quantitative numbers, but added nothing that was very useful. We still have those tapes. If anybody would like to pursue this project, you are welcome to them, but I have not done any mathematical analysis of these functions for at least four years. Our clinical experience, of that time, at least 100 patients, was published in *Radiology*, about three years ago. The title is something like: Temporal Pathological